Anchors—Tests Procedures and Vibratory Analysis

Jean Jacques Rincent

Anchors—Tests Procedures and Vibratory Analysis

Springer

Jean Jacques Rincent
Évry-Courcouronnes, France

ISBN 978-981-96-3779-9 ISBN 978-981-96-3777-5 (eBook)
https://doi.org/10.1007/978-981-96-3777-5

This Springer imprint is published by the registered company Springer Nature Singapore Pte Ltd.
The registered company address is: 152 Beach Road, #21-01/04 Gateway East, Singapore 189721, Singapore

Preface

This book follows: Ground Anchors: Tension Force Vibratory Analysis JJ. Rincent. Springer Sept 2024.

This new book is a summary of experience gained from the analysis of over 24,000 test curves obtained from 3000 ground anchors analysis. It provides practical feedback on the complexity of non-destructive testing. The test equipment, the rules for acquiring understandable results. The examples come from trials carried out in Brazil over the last 5 years and finally the test methodology adopted for this experiment.

These tests are generally carried out on ground anchors using strands as reinforcement or rebars several decades old. Information on their length and initial tension force is often non-existent after 40 years. That's why these vibration analysis tests provide the answers and information needed for stability and maintenance diagnostics.

The examples chosen concern ground anchors equipped with bars, strands and also for nails, passive ground anchors and micro piles. An example on anchors of a concrete dam is given. Tests on prestressed dowels are described, and a test on a reinforcing bar linked to the construction of an early-sixteenth century castle will be carried out.

As a reminder that these tests results lead to define the total length of the tie bar, the free length, the diameter of the tie bar, i.e. the reinforcement with its cement grout, and finally the tension force at the time of testing. These data are essential for assessing the stability of retaining walls stabilised by tie rods.

It should be stated that static tests, which are difficult to carry out at height, provide no information on lengths and run the risk of breaking old tie rods.

All the tests carried out are used to construct the test method that can be adopted, taking into account the feedback acquired from several hundred static tests and several thousand dynamic tests.

Évry-Courcouronnes, France　　　　Jean Jacques Rincent

Competing Interests The author has no competing interests to declare that are relevant to the content of this manuscript.

Introduction

For retaining structures and tie rods in particular, access to the head of the tie rods must be preserved in order to:

- Carry out inspection tests.
- Re-tension the tie rods, if necessary.
- While protecting them from corrosion.

Meeting these conditions means implementing sustainable maintenance of the structures. This test method is a diagnostic tool for ground anchors used by managers of retaining structures to design maintenance projects. The final aim is to increase the durability of the retaining structures.

The tests proposed and explained using numerous examples and finally to propose a methodology for carrying out vibration analysis tests on ground anchors, as well as a framework for their interpretation.

Contents

1 Method and Equipment 1
- 1.1 Introduction 1
- 1.2 Theory 1
- 1.3 Test Equipment 3
- 1.4 Tie Rod Head Protection 4
- 1.5 Location of the Test Itself 7
- 1.6 Tests Sampling 7
- 1.7 Types of Ground Anchors 11

2 Dynamic Tests 13
- 2.1 Standard Curves 13
- 2.2 Acquisition Frequencies 13
- 2.3 Number of Impacts and Curves 15
- 2.4 Results Analysis 17
- 2.5 Low Tension Forces 19
- 2.6 Concrete Quality 20
- 2.7 Hidden Part of the Head 22

3 Static and Dynamic Tests—Simultaneous Tests 25
- 3.1 Static Test's Location 25
- 3.2 Static Tests—Concrete Dam 27
- 3.3 Dynamic Tests 29
- 3.4 Simultaneous Static and Dynamic Tests 30

4 Presence of Bulbs of Cement Grout 35
- 4.1 Equivalent Cross-Section Calculation 35
- 4.2 Static Test and Bulb of Cement Grout 35
- 4.3 Others Examples 38

5 Dynamic Stiffness/Static Stiffness ... 43
5.1 Dynamic Stiffness ... 43
5.2 Static Stiffness ... 43
5.3 Dynamic and Static Tests ... 44
5.4 Dynamic Stiffness/Static Stiffness ... 47

6 Tension Force Calculation ... 49
6.1 Common Case ... 49
6.2 Additional Analyses ... 50
6.3 Inertia Combination ... 51
6.4 Curves Interpretation—Stiffnesses ... 53

7 Curves Response—Analysis ... 55
7.1 Dynamic Stiffness—Relative Uncertainty ... 55
7.2 Vibration Response During the Static Test ... 56
7.3 Static Test—Unloading ... 60

8 Uses of the Static Tests ... 63
8.1 Before Static Tests ... 63
8.2 Results Analyses ... 63
8.3 Deformation Rules ... 65
8.4 Equipment to Perform the Tests ... 67

9 Repro-tension ... 69
9.1 Electrical Isolations ... 69
9.2 Interest of Repro Tension ... 70
9.3 Procedure ... 70
9.4 Plate Thickness Calculation ... 70
9.5 Particular Cases ... 71
9.6 Durability ... 72

10 Pre-tensioned Dowels ... 75
10.1 Laboratory Tests ... 75
10.2 Pre-tensioned Bolts ... 75
10.3 Pylons Fixations ... 77
10.4 Floor Mounting of Industrial Robots ... 79

11 Tension Force Limit of an Element Fixed to a Support ... 83
11.1 Tests Conditions ... 83
11.2 Procedures ... 83

12 Micro Piles ... 89
12.1 Current Tests ... 89
12.2 Soil Skin Friction Scaling ... 89

13 Other Applications 95
13.1 Soils Nailing 95
13.2 Embankment Reinforcement 97
13.3 Historic Monuments 100
13.4 Data Base 100

14 Test Procedure and Results 105
14.1 Dynamic Test 105
14.1.1 Definitions 105
14.1.2 Results 106
14.1.3 Elements of the Acquisition Chain 106
14.1.4 Procedure 107
14.1.5 Results 108
14.1.6 Test Report 109
14.1.7 Verification of the Equipment 109
14.2 Static Tests 110
14.2.1 Particular Cases 110
14.2.2 Simultaneous Static and Dynamic Tests 110
14.2.3 Results 110
14.3 Final Objectives 111

15 Conclusions 115

Appendix A 117

Appendix B 119

Appendix C 125

Appendix D 127

Appendix E 131

Appendix F 135

Appendix G 137

Appendix H 139

Appendix I 141

Appendix J 143

Appendix K 145

Appendix L 147

Appendix M 149

Appendix N 155

Bibliography 157

About the Author

Dr. Jean Jacques Rincent Doctor of Civil Engineering (Institute National des Sciences Appliquées of Toulouse INSA) President and founder of Rincent BTP Services.

I was one of the administrators of the National Agency for Research and Technology ANRT from 2017 to 2023.

COFRAC (2011–2019) technical expert Comité Français d' Accréditation specialities:

- Geotechnical testing—laboratory and in situ.
- Instrumentation and monitoring of structures.
- Earthworks.

Member of the standardisation committee: Sols Reconnaissance Essais. Participation in the drafting of 65 standards.
The main lines of development are:

- Determination of the anchors tension force with vibratory analysis.
- Dynamic Behaviour of Pavement under Impact Loading.
- Development of a procedure for analysing the bitumen content of asphalt mixes with the Ground Penetration Radar.

Former Director of the Geomechanics Department of the CEBTP (Experimental Centre for Building and Public Works) for 10 years.
Associate Assistant Professor Polytech Lille Civil Engineering section for 2 years and at the start teacher at the Civil Engineering Department of the National Engineering School of LIBREVILLE Gabon, School linked by contract to INSA (Institut National des Sciences Appliquées) of Toulouse.

Chapter 1
Method and Equipment

Abstract After a brief review of the theory behind the test, the equipment used is presented. The practical implementation of the test is described, including the location of the sensors and the difficulties in accessing the heads of the old tie rods.

1.1 Introduction

The vibration analysis method used to analyse ground anchors was designed at the French Building Research Centre (CEBTP) in 1974 and used to the anchors since 2000. The original name was" méthode impulsionnelle» or impulse method.

This method, devised by J. Paquet and implemented in 1960, was formalised in 1968. Allan Davis then wrote a number of articles on the possible uses of this technique in civil engineering works.

Before the year 2000, we developed the application linked to tie rods and prestressed dowels. At the end of 2018, we began to apply our know-how to a number of tensioned walls in Brazil. This book gathers our experience in the field through cases studied.

Remember that this dynamic method can be used to diagnose tie rods and determine their internal tension without causing them to break. In the case of old tie rods, the real risk exists with static testing.

Last but not least, the means used for dynamic testing make it possible to carry out tests at height.

1.2 Theory

Acquisitions are made in temporal mode and curves are used in frequential mode (Fig. 1.1).

The following curve is obtained:

This curve leads to calculate:

J. J. Rincent, *Anchors—Tests Procedures and Vibratory Analysis*,
https://doi.org/10.1007/978-981-96-3777-5_1

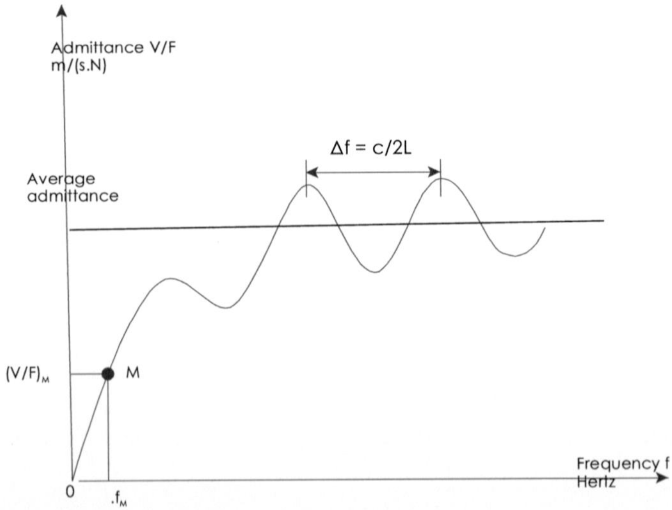

Fig. 1.1 V/F function of the frequency

- the total anchor length,
- the free length,
- the cross section of the anchor with its grout
- and the dynamic stiffness.

It is necessary to measure the force of the impact generated by the hammer, which is equipped with a piezoelectric sensor.

The general scheme is given in the appendix A, extract from the European patent n° (11) EP 1 761 750 B1. Theoretical elements are given in the attached list of bibliographical references.

The content of this book focuses on the calculation of dynamic stiffness and their use to calculate tension forces.

Dynamic stiffness is a complex number. Research has shown that the real and imaginary parts are constant for values below 80 Hz. In our case tie-rod stiffness is calculated for frequency values below 40 Hz.

There is a relationship between the dynamic stiffness R_d and the force F for the tie rods which is of the following form:

$$(R_d)1/2 = aF + b$$

We therefore need to determine a and b to establish the formula linking dynamic stiffnesses and tension forces.

The system described below enables dynamic and static tests to be carried out simultaneously on the same element. This is one way of calculating a and b.

1.3 Test Equipment

The central element of the acquisition process is a microcomputer equipped with a battery for acquiring test data over the course of a day.

The hammer has to be a well-balanced object, because we mustn't forget that the majority of tests are carried out at height, which is one of the reasons why this method is used (Fig. 1.2).

The surface to be impacted by the hammer must be clean and smooth (Fig. 1.3).

The tip of the hammer must strike flat against the surface of the head of the element under test. The sound of the impact is characteristic, and an experienced operator can quickly tell whether the shock emitted is of good quality.

Fig. 1.2 Test conditions. *Source* Rincent BTP Recife

Fig. 1.3 Grinding the test surface. *Source* Rincent BTP Recife

Fixing the 3D geophone to the plate at the head of the tie-rod is an essential part of the test acquisition process. This fixation must be rigid enough to measure the dynamic stiffness of the tie-rod. The photos show a working example of this type of fixing. Once again, this point is essential if you want to obtain usable results (Figs. 1.4 and 1.5).

Here, the solution is to glue a small block of bakelite or equivalent onto which the geophone's metal plate is screwed.

Finally, an important point must be made to ensure that the hammer does not strike and crush the connection cable of the hammer.

1.4 Tie Rod Head Protection

All the examples given here come from work carried out in Brazil, where the heads of the tie rods are protected using concrete. The first question that is usually asked is whether tests can be carried out on these concrete protections. We have carried

Fig. 1.4 Geophone fixation

Fig. 1.5 Geophone fixation. *Source* Rincent BTP Recife

out a number of tests, which confirmed that this method introduces a significant uncertainty factor. This concrete protection is destroyed to gain access to the head of the tie rod (Figs. 1.6 and 1.7).

Fig. 1.6 Concrete blocks protections

Fig. 1.7 Destruction before tests. *Source* Rincent BTP Recife

Tests were carried out on three tie rods before and after the destruction of the concrete protection head. In all three cases, the dynamic stiffness measured was higher when the protective head was removed. The results are very similar for one of the tie rods and very different for the other. The presence of concrete at the head introduces a parameter of uncertainty that must be avoided by accessing the head of the tie directly (Fig. 1.8).

The real issue is to recommend the installation of a removable protection system that allows static or dynamic tests to be carried out easily, while ensuring protection against corrosion. Access to the head of the tie rod is a problem both during testing and when the tie rod needs to be re-tensioned. Our Brazilian experience with various structures and over 3000 ground anchors shows that the annual loss of tension varies from 1 to 5%. It is therefore necessary to access the heads of the tie rods to ensure their maintenance and the stability of the structures.

In the following photo, the tie-head protection device enabled to carry out vibration analysis tests (Fig. 1.9).

1.5 Location of the Test Itself

The very first tests were clarification tests positioning of the 3D geophone and impacts of the instrumented hammer to measure the force. For example, the following photos show two possibilities for carrying out dynamic tests. The aim is to calculate the tension force of the tie-rod (Figs. 1.10, 1.11 and 1.12).

Strikes A to E are made on the metal plate and strikes F to J are made on the tie-rod reinforcement. Without going into too much detail, and when compared with the results of the static tests, the plate method appeared to be very reliable for the purpose of determining the internal tension of the tie-rod. Remember that the number of static tests carried out to calibrate the dynamic method is significant. For example, on a site where 460 dynamic tests were carried out, on 9 contention walls, 22 static tests were carried out i.e. more of 4% of the tested tie rod.

1.6 Tests Sampling

When analysing test results, we must take into account the characteristics of the test location, whether static or dynamic. Testing in the proximity of a construction joint or drainage structure may have an impact on the results obtained. Cracking, poor contact between the distribution plate and the supporting wall, etc., should be noted, photographed and archived. This is particularly true for static tests, where the test equipment is heavy and difficult to handle at height. When using the results, therefore, this particular location must be taken into account, as it does not generally represent the most common cases.

222			231			240		
dynamic	1,12E+09	1,31E+09		2,23E+09	3,39E+08		1,98E+09	1,75E+09
stiffnesses	1,26E+09	1,26E+09		2,19E+09	8,89E+08		1,98E+09	1,72E+09
N/m	1,40E+09	1,26E+09		2,20E+09	2,11E+08		2,00E+09	1,72E+09
	1,51E+09	1,34E+09		2,18E+09	1,54E+09		1,85E+09	1,71E+09
	1,34E+09	1,26E+09		2,21E+09	1,58E+08		2,02E+09	1,76E+09
	1,17E+09			2,29E+09			1,89E+09	
average	1,30E+09	1,29E+09		2,22E+09	6,27E+08		1,95E+09	1,73E+09

Fig. 1.8 Stiffnesses results N/m

Fig. 1.9 Access to anchor head. *Source* Rincent BTP France

Fig. 1.10 Strike on the plate

In the figure below, which shows a test carried out at the foot of the retaining wall, the inertia of the reinforced concrete structure must be taken into account, as it carries away run-off water and contributes to the stability of the wall. This test is not representative of a current part of the structure (Fig. 1.13).

Concerning the number of elements to be tested, if the test campaign is informative, the number of anchor tests may be limited. However, for the sampling to be representative of the tie rods, it is first necessary to identify families of tie rods by line, by location, e.g. close to a construction joint, at the ends of the wall where backfill volumes decrease, etc.... A percentage of tested tie rods of less than 20% should be avoided.

Fig. 1.11 Strike on the anchor head. *Source* Rincent BTP Recife

106	A	9,38E+07		116	A	1,20E+09		126	A	1,41E+09		135	A	1,79E+09	
	B	2,23E+08			B	1,26E+09			B	1,38E+09			B	1,68E+09	
	C	2,66E+08			C	1,34E+09			C	1,39E+09			C	1,76E+09	
	D	2,41E+08			D	1,21E+09			D	1,38E+09			D	1,64E+09	
	E	3,30E+08	2,31E+08		E	1,30E+09	1,26E+09		E	1,45E+09	1,40E+09		E	1,60E+09	1,69E+09
	F	5,99E+07			F	1,14E+09			F	1,25E+09			F	1,59E+09	
	G	2,22E+08			G	1,19E+09			G	1,27E+09			G	1,53E+09	
	H	7,57E+07			H	1,19E+09			H	1,21E+09			H	1,57E+09	
	I	5,56E+08			I	1,16E+09			I	1,23E+09			I	1,43E+09	
	J	5,49E+08	2,98E+08		J	1,21E+09	1,18E+09		J	1,25E+09	1,24E+09		J	1,40E+09	1,50E+09

Fig. 1.12 Results table Stiffnesses N/m

Fig. 1.13 Tests near structure of concrete. *Source* Rincent BTP Recife

When the aim is to obtain information useful for maintenance, it is preferable to test all the tie rods. Stability calculations will be carried out using all the information provided by each of the tie rods.

The number of static tests performed must be sufficient to calibrate the vibration analysis tests. It is understood that, for example in the following figure, it is necessary to calibrate the tie rods located on the slope of the embankment and those involved in supporting the vertical wall (Fig. 1.14).

Fig. 1.14 On the same site, two types of anchors. *Source* Rincent BTP Recife

It will also be necessary to make a distinction for bars of different diameters or for tie rods with 8, 10 or 12 strands.

The number of static tests varies from 1 to 3% depending on the structure, the type of tie rods and the number of tie rods tested.

These tests results are used to:

- know the loss of tension
- identify the anchors which will be re-tensioned
- and to elaborate the procedure of re-tensioning.

1.7 Types of Ground Anchors

It should be borne in mind that the tie rods tested are between 30 and 45 years old and that, at the time they were tested, the types of tie rods were very diverse. Bars of all diameters, trends and anchors with flat jack were encountered on the sites treated, which meant that the means had to be adapted, especially for the static tests.

The 6 examples in the figures below show the difficulty of adapting the equipment used to carry out the tests (Fig. 1.15).

Fig. 1.15 Diferent types of ground anchors encountered. *Source* Rincent BTP Recife

Chapter 2
Dynamic Tests

Abstract The standard curves are described, frequencies, number of impacts. Analyses examples are given and particular cases of ground anchors with a low-tension force is given. The influence of the concrete quality and of cavity under the iron plate located at the tie rod head are presented.

2.1 Standard Curves

The force signal in blue and the velocity signal in red are shown as a function of time (Fig. 2.1).

Note that a tie-rod vibrates strongly, and after applying the Fourier transforms, the V/F curve as a function of frequency can be used to calculate the dynamic stiffness.

When tests are being carried out, it is necessary to identify the events that may produce vibrations that will add to the vibration response of the tie rod. According to the response obtained on a tie rod located on a dam, the first response is obtained in 'calm' period, the second is obtained when the spillway is operating (Figs. 2.2 and2.3).

2.2 Acquisition Frequencies

Acquisition frequencies can be different depending on the use of the V/F curve as a function of frequency. For dynamic stiffness it is not useful to have a high frequency range, similarly for identifying the thickness of the retaining wall and the often-overabundant grout immediately adjacent to the wall in the ground. If the chosen frequency is less than 800 Hz the wall response cannot be see.

The figure below shows the response curve obtained in dark blue and the light blue curve obtained by removing the response of the retaining wall with the grout bonded to it. The overall response is the geometric sum of the individual responses (Figs. 2.4, 2.5 and 2.6).

J. J. Rincent, *Anchors—Tests Procedures and Vibratory Analysis*,
https://doi.org/10.1007/978-981-96-3777-5_2

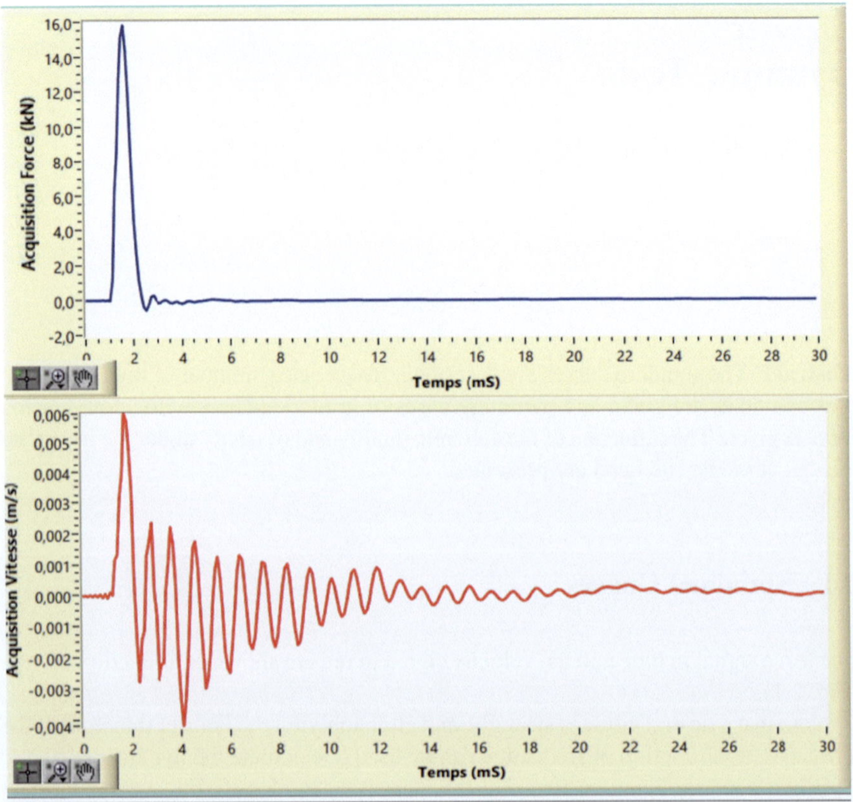

Fig. 2.1 Force and velocity responses

Appear the following elements: the total length, the free length, the thickness of the retaining wall plus, in general, the grout bonded to the wall, the tie rod diameter and the dynamic stiffness.

For calculating lengths, the velocity used is 4000 m/s, but experience shows that the real velocity in a tie is higher than 4000 m/s and often close to 4500 m/s.

The frequency ranges chosen for acquisitions are usually 1300 and 1800 Hz.

The cross-section calculation is done with the average mobility (Fig. 2.7).

Mobility as a function of the cross section of the tested element

$$V/F = 1/\rho_b V_b A$$
$$V/F\,\mathrm{m/sN}$$

Wave velocity in the tie rod V_b m/s.
Concrete volume mass ρ_b in kg/m^3
Cross section area A (m^2).

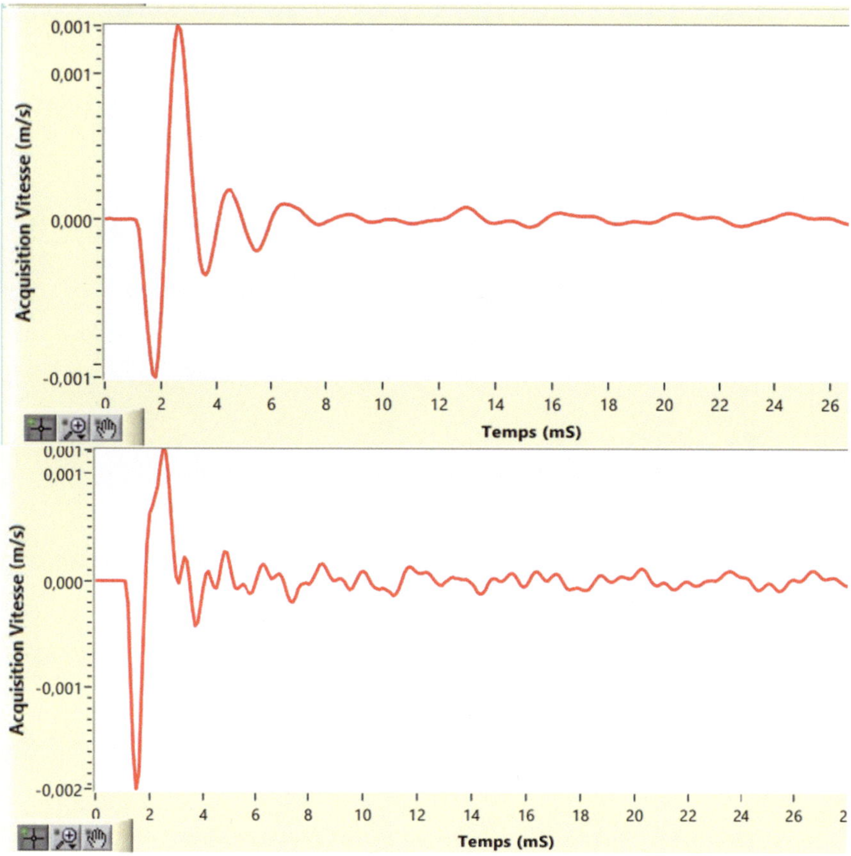

Fig. 2.2 **a** Calm period. **b** Spillway operating

2.3 Number of Impacts and Curves

At the first beginning years the test number was 6, but rapidly it appears that two values were very often out of the average. Now 8 tests are carried out for a non-destructive test. It is interesting to carry out these acquisitions in order to eliminate the highest and lowest dynamic stiffness values. Very often the first value of the dynamic stiffness is high, followed by a low value. The compression wave caused by the hammer generates changes in the behaviour of the tie rod. The tests are also carried out in different frequency ranges to better identify the vibratory responses of free and total lengths (Fig. 2.8).

The table in the following figure shows two columns of results from the ground anchors on a retaining wall with more than 700 tie rods. These anchors are located in the central part of the wall, which has 12 lines of tie rods (Appendix B) (Fig. 2.9).

Fig. 2.3 Spillway view. *Source* Rincent BTP Recife

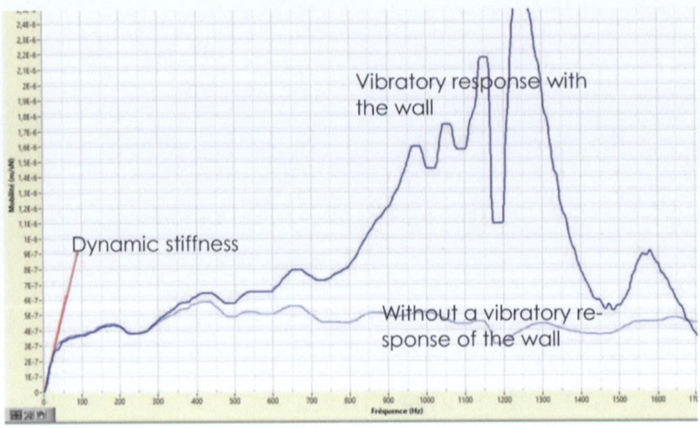

Fig. 2.4 Curves responses

Fig. 2.5 Classical curve—length of the tested element

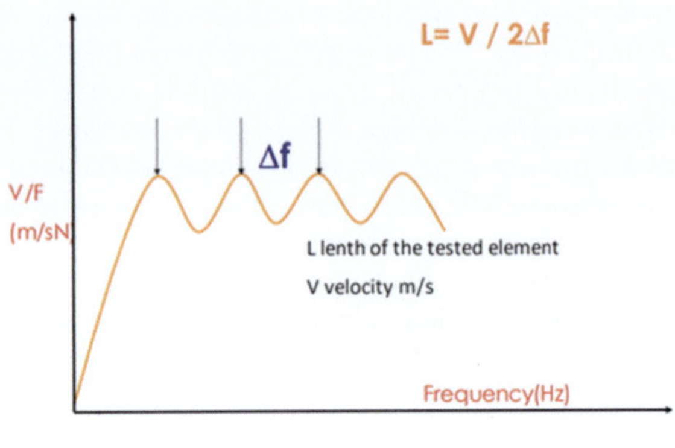

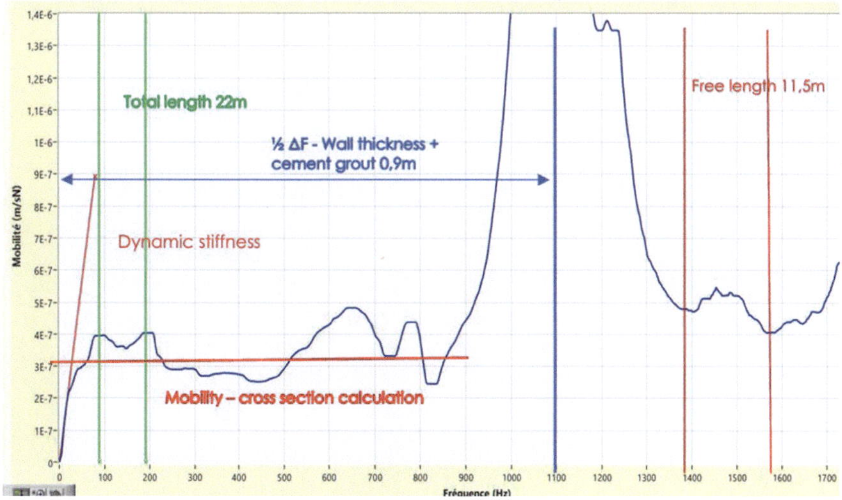

Fig. 2.6 Curve obtained for a ground anchor

Fig. 2.7 Average mobility

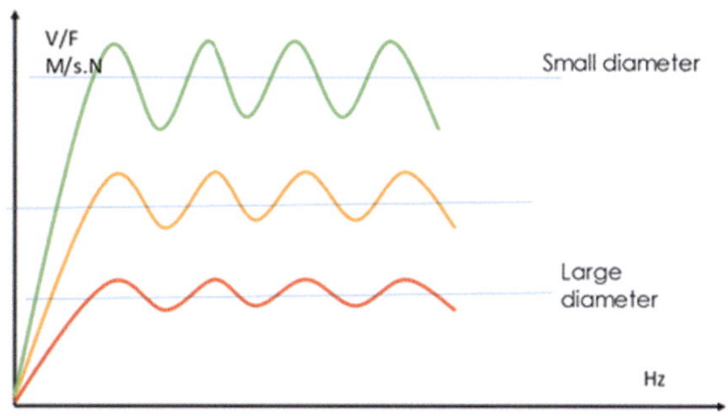

Four families of results can be identified based on the average dynamic stiffnesses and for tension forces from 0 to 5 tons from 5 to 10 tons from 10 to 20 tons and over 20 tons. All the dynamic stiffness results for these two columns of anchors are gathered in Appendix B.

2.4 Results Analysis

In the example analysed, the anchors multi-strand are 40 years old. The table below shows the 5 dynamic stiffness values used to calculate the average and then the relative uncertainty of the stiffnesses values of tie rods with an internal tension greater than 20 tons (Fig. 2.10).

Fig. 2.8 On site tests. *Source* Rincent BTP Recife

Dynamic stiffnesses N/m

Line number		
1	5,40E+08	4,70E+08
2	1,14E+09	5,63E+08
3	1,58E+09	5,80E+08
4	1,86E+08	1,12E+08
5	1,49E+08	2,01E+08
6	1,55E+08	2,09E+08
7	2,86E+09	3,56E+08
8	1,02E+08	1,68E+08
9	1,41E+08	1,98E+09
10	1,87E+09	2,86E+08
11	3,12E+09	2,62E+09
12	2,72E+09	2,19E+09

Force inferior to 5 tons

5 to 10 tons

10 to 20 tons

superior to 20 tons

Fig. 2.9 Dynamic stiffnesses values, central part of the wall

2,73E+09	3,11E+09	2,73E+09	1,95E+09	1,96E+09
2,91E+09	3,02E+09	2,75E+09	2,22E+09	2,62E+09
2,76E+09	3,22E+09	2,69E+09	2,32E+09	2,65E+09
2,60E+09	3,24E+09	2,73E+09	2,35E+09	2,68E+09
3,11E+09	3,23E+09	2,73E+09	2,39E+09	2,93E+09
3,02E+09	2,92E+09	2,71E+09	1,93E+09	2,87E+09
6,71%	4,23%	0,76%	9,32%	13,21%

Fig. 2.10 Relative uncertainly of the stiffnesses

The relative uncertainty is used to quantify the sensitivity of the tie rod to the compression waves generated by the impact of the hammer. It is also a means of assessing the stability of the tie rod during re-tensioning.

9 out of 24 tests, i.e. 40% of the dynamic stiffness values, give a higher stiffness on the first strike because the first strike incorporates a system stiffness that includes, for example, corrosion-related links that disappear or are reduced on subsequent strikes. This is why the number of eight strikes for a test is a way of eliminating extreme stiffness values.

In 5 cases out of 24, i.e. 20%, the lowest stiffness value follows the highest stiffness value, which corroborates the phenomena observed.

In conclusion, for this family of 40-year-old tie rods, the internal tension of at least 20 tons means that they are not very sensitive to impacts from the hammer and can be potentially re-tensioned.

2.5 Low Tension Forces

The four tie rods that lost the most tension have internal tensions of less than 5 tons and relative uncertainties that vary from 15 to 50%. The first phase consists of combining static tests with tie rods identified as representative of families of tie rods and putting them under tension using a combination of static tests.

The re-tensioning of the tie-rod with a relative uncertainty of 15% will be possible, but re-tensioning the tie-rod with a relative uncertainty of 50% seems more hypothetical (Figs. 2.11 and 2.12).

The dynamic stiffnesses of the tie rods where the tension is less than 5 tons are shown in the attached graph. Each hammer strike corresponds to the value of the stiffness, it being understood that the highest and lowest of these values have been removed. The six remaining values are used to understand the behaviour of each of them.

For the number of hits, we systematically perform 8 hits, which allows us to analyse the six remaining hits after removing the extreme minimum and maximum values. Our concern is not to penalise the test by increasing the difficulty of carrying it out, while at the same time recovering the elements necessary for a good analysis.

1,53E+08	1,59E+08	1,54E+08	1,19E+08
1,85E+08	1,03E+08	2,10E+08	1,88E+08
1,67E+08	1,29E+08	1,13E+08	1,95E+08
2,15E+08	8,40E+07	2,21E+08	5,06E+08
1,73E+08	9,28E+07	2,45E+08	3,04E+08
2,22E+08	1,06E+08	2,63E+08	4,01E+08
14,73%	24,46%	28,17%	51,40%

Fig. 2.11 Relative uncertainly of dynamic stiffnesses

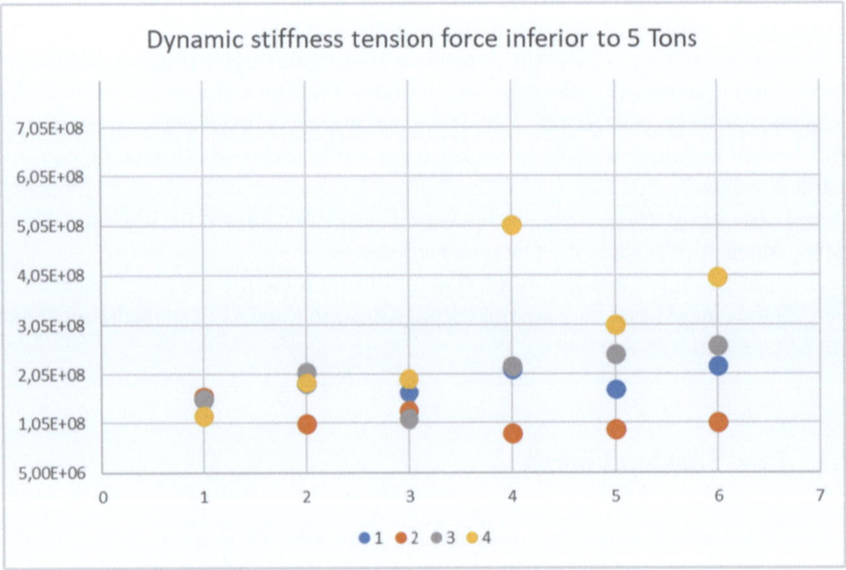

Fig. 2.12 Behaviour of the tie rod under hammer impacts

2.6 Concrete Quality

The concrete quality of the support structures is generally good, even after more than 40 years. However, the example presented below shows that in this case, the addition of a coating layer induces a weak point that quickly became apparent during dynamic and static tests (Fig. 2.13).

For the dynamic tests of the T44 are very specific to concrete with poor surface quality, dynnamic stiffness (N /m) function the hammer blow number:

The dynamic stiffness decreases from 2.67E8 N/m to 9.35E5 N/m, the quality of this coating is so poor that blows from a hammer on the tie-rod support plate have destroyed it.

Fig. 2.13 Coating layer. *Source* Rincent BTP Recife

2.7 Hidden Part of the Head

The influence of the hidden part of the element must be detected when the result analysis is done (Fig. 2.14).

These flat cylinders were used to finalise the tensioning of the tie-rod. When the required tension was reached, the resin-filled cylinder was left as it was. Here you can see unfilled spaces under the metal cylinder. Depending on the position of the 3D geophone and the strikes, the dynamic stiffness values can be very different (Fig. 2.15).

This tie rod has been retested and after that, the head investigation has been done. The operator usually detects the presence of voids or an anomaly by the sound of the hammer striking the metallic plate.

Fig. 2.14 Under the metallic plate. *Source* Rincent BTP Recife

Fig. 2.15 Dynamic stiffnesses N/m tie rod 9

9	9,01E+07	re tested	9	1,10E+06	
	3,68E+08			4,84E+07	
	3,95E+08			1,76E+08	
	4,14E+08			1,91E+08	
	4,22E+08			2,00E+08	
	3,48E+08			2,23E+08	
	4,03E+08			4,30E+07	
	3,95E+08	3,87E+08		1,89E+08	1,41E+08

Chapter 3
Static and Dynamic Tests—Simultaneous Tests

Abstract The influence of the location of the tests is shown on a test at the foot of a retaining wall where there is a concrete rainwater drainage structure. The example of simultaneous tests is given for the ground anchors of a concrete dam, with interpretation.

To establish a formula linking dynamic stiffness and force for old and sometimes very old tie rods, it is necessary to carry out static tests to find out the tension force of the tie rod. It is sometimes possible to obtain information on site when tie rods are fitted with measuring cells. Tie rods with zero tension are also sources of information.

3.1 Static Test's Location

The rainwater drainage structure contributes to the stability of the wall and absorb part of the forces. The extract from the table showing the forces in the tie rods at the base of the wall shows the influence of the concrete structure Appendix C (Figs. 3.1 and 3.2).

The forces generated by the ground pressure increase with the height of the embankment behind the retaining wall. Here the forces are reduced by the presence of the rainwater drainage structure.

This is why, whenever possible, tests should exclude ground anchors located close to concrete structures that evacuate rainwater.

On another point, experience gained from re-tensioning tie rods shows that the internal tension of the tie rod determined from the loading phase overestimates the tension force by around 1 tonne, if the value is taken during the loading phase, it is better to use it during the unloading phase. For the old anchors, adhesive forces exist for example the corrosion, the plate contact with the concrete wall.

J. J. Rincent, *Anchors—Tests Procedures and Vibratory Analysis*,
https://doi.org/10.1007/978-981-96-3777-5_3

Fig. 3.1 Static test location. *Source* Rincent BTP Recife

Fig. 3.2 Force de tension at the bottom of the wall

T227	19,0t	↗	T239	15,7t	↗
T228	17,5t	↘	T240	12,8t	↘

3.2 Static Tests—Concrete Dam

The first step is to estimate the resources required to carry out the static tensile test on the tie rod. In the case of old tie rods, it is advisable to obtain information on the type of steel used in the reinforcement. We will take the example of tie rods made up of strands ensuring the stability of a dam. It should be borne in mind that carrying out a static test could lead to the failure of the tie rod, which is a risk since the tie rod contributes to the stability of the structure (Figs. 3.3 and 3.4).

The main difficulty in carrying out a tensile test is mechanical—how to connect the head of the tie rod when it is not designed for this operation, especially on old tie rods. When the internal forces are less than 45–50 tonnes, this is a difficulty that can be resolved by experienced operators using simple means. For higher forces, this operation quickly becomes complicated and even impossible to carry out with simple means.

For old tie rods, in this case 35 years old, the first step is a literature search to find out the type of steel used and to take a sample to carry out a direct tensile test. The results of these tests enabled us to estimate the initial tension force at 100 tonnes. The results of various tests carried out in Brazil on tie rods of the same type suggest tension losses of 1–2% per year in similar configurations.

Assuming 2%, the force value would be 49.3 tons, while assuming 1%, it would be 70 tons after 35 years.

This approach makes it possible to define the means to be used to carry out the tests:

- Connection of the head to the traction device
- Hydraulic means for carrying out the tests
- Mechanical dimensioning of the device to carry out simultaneous static and dynamic tests.

Some tie rods were instrumented at the time of the works, but after 35 years they no longer function. It was therefore necessary to calibrate by means of static tensile tests to determine the values a and b of the formula relating the square root of the stiffness and the internal tension force of the tie rod.

$$(\mathrm{Rd})1/2 = \mathrm{aF} + \mathrm{b}$$

It is understood that on certain dams, information exists on the tension forces of identified tie rods, which can be used by carrying out dynamic tests on these tie rods.

Fig. 3.3 Anchor drawing

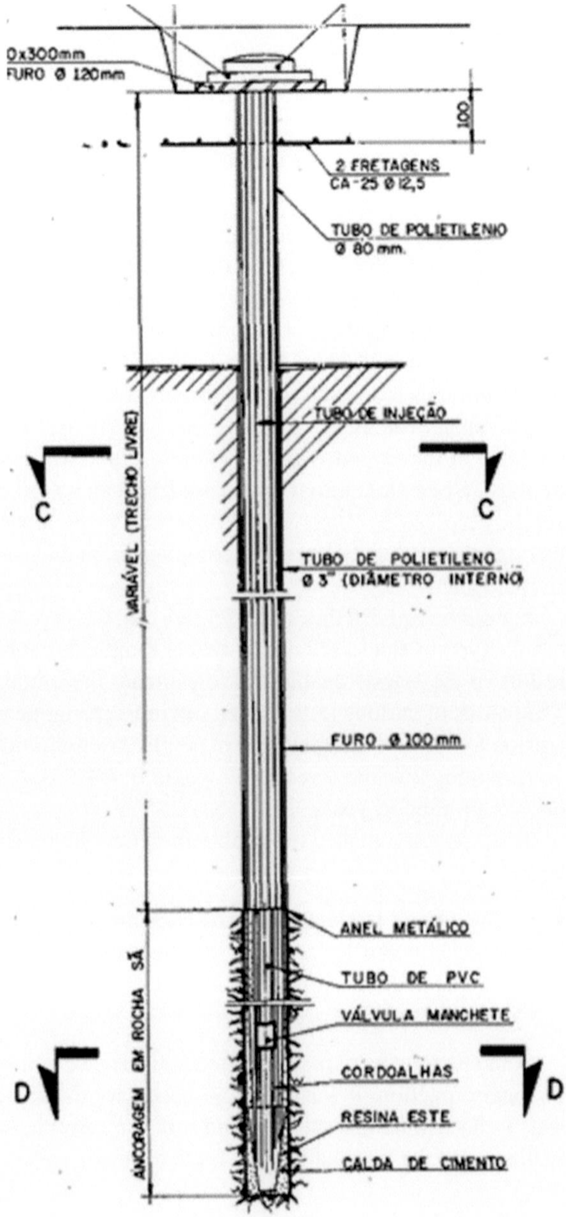

Fig. 3.4 Concrete dam and anchors head. *Source* Rincent BTP Recife

3.3 Dynamic Tests

In order to carry out simultaneous static and dynamic tests, a metal structure needs to be designed which can:

- to be fixed to the head of the tie rod or by connecting the strands of the tie rod with strands that will ensure transmission of the force
- while allowing non-destructive tests to be carried out at each bearing load level (Fig. 3.5).

Tie rod 42 was first tested dynamically to obtain results under the existing initial site conditions. It was then tested simultaneously in static and dynamic conditions.

This procedure eliminates any disturbances that may be generated by the installation of the tensioning device.

The static tensile test was carried out as follows:

- Loading stages 1.2t 10t 20t 30t 40t 50t 55t
- Unloading idem

Fig. 3.5 Test devices. *Source* Rincent BTP Recife

- Duration of stops, loading 15 min
- Unloading 5 min.

It should be borne in mind that a tensile test carried out on a 35-year-old tie rod presents a risk and that the procedure to be implemented has nothing to do with, for example, a procedure for accepting a new tie rod for validation.

On one site, the tensile test was stopped because the operator could hear the strands breaking when the tension was increased by changing the load level. The tensile test must take into account the fact that the ground anchor under test is old and may be in a condition close to failure. This observation implies having experienced operators on site who can decide on their own to stop a test.

3.4 Simultaneous Static and Dynamic Tests

The example chosen, concerning the determination of the internal tension force of the tie-rod 253, shows the concordance of the operator's observations on site with the measured dynamic results, which is not the most common case. Values measured on site during the loading of the tie-rod lead to an overestimation of this internal force in the tie-rod (Figs. 3.6 and 3.7).

Fig. 3.6 Simultaneous tests. *Source* Rincent BTP Recife

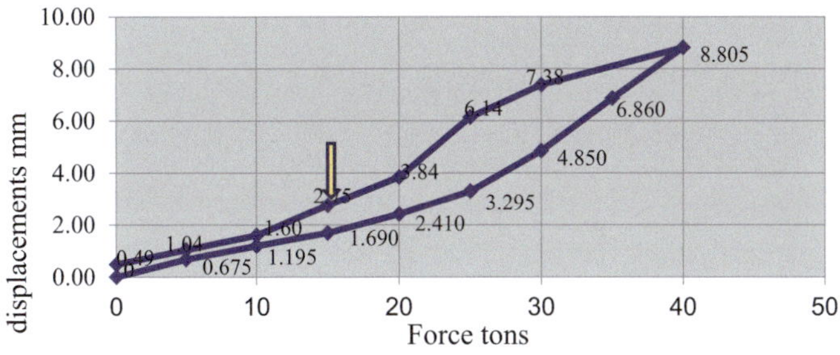

Fig. 3.7 T 253 Static test

In this case, the operator indicated that the metal plate had lifted off the wall at a force of 15 tons.

Dynamic tests are carried out at each loading and unloading stage. The first gives all the continuous dynamic stiffness values and identifies 15 tons as the internal force of the tie rod, which is also shown in the average values calculated per stage.

Hammer strike number, 1–8, 1 ton; 9–16, 5 tons; 17–24, 10 tons; 25–32, 15 tons. (Appendix D)

In this example, the static test leads to the same conclusion, the tension force is 15 tons (Figs. 3.8 and 3.9).

Details of the metal structure used for these tests help to understand the variations in dynamic stiffnesses (Figs. 3.10 and 3.11).

The sudden change in dynamic stiffness values corresponds to a change in load application when the tensile force on the tie rod exceeds the internal tension force. The application of the tensile test force passes from the distribution plate located at the head of the tie rod in position 1 to the two metal supports in position 2.

The main purpose of the static and dynamic tests carried out simultaneously is to determine the internal tension force of the tie-rod and to use this result to calibrate

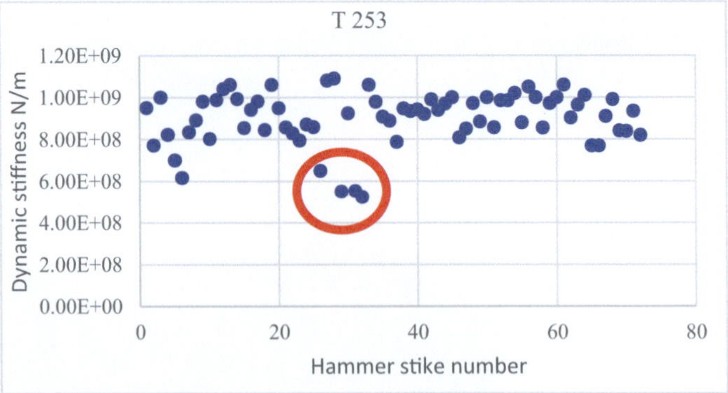

Fig. 3.8 Dynamic stiffness results

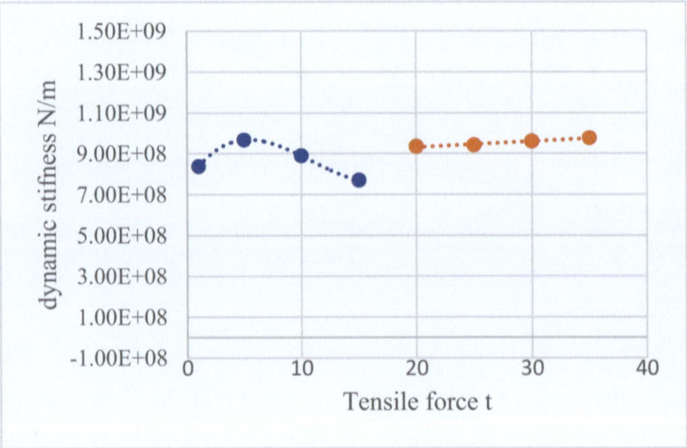

Fig. 3.9 Average stiffnesses T 253

the dynamic tests. Both types of tests provide further information on the behaviour of the tie rod in terms of its use.

Fig. 3.10 Plate detachment. *Source* Rincent BTP Recife

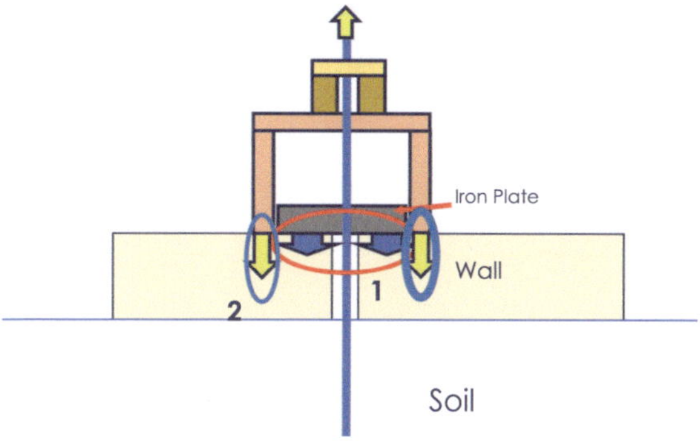

Fig. 3.11 Scheme of the traction device

Chapter 4
Presence of Bulbs of Cement Grout

Abstract Sometimes a bulb of cement grout exists at the contact of the wall in the ground. The dimensions of this bulb have an effect on the results of both the static and dynamic tests described in this chapter.

4.1 Equivalent Cross-Section Calculation

Mechanical admittance V/F (m/sN) is a function of the cross-section A of the tested element

$$V/F = 1/\rho_b V_b A$$

Wave velocity in the tie rod V_b m/s, if no value measured, take 4000 m/s.

Concrete volume mass ρ_b in kg/m^3 and here take 2400 kg/m^3

For the tie rod of the Fig. 4.1, the cross-section area A (m^2) is 0.10 m^2, i.e. an equivalent diameter of 36 cm. It is understood that, in the case of tie rods, the cross-section of the borehole is generally oval and more developed towards the bottom given the physical conditions in which it is made.

This curve was analysed following the result of the static test presented in the next paragraph. The equivalent diameter calculated for this vibratory response is 65 cm, i.e. 80% greater than the diameter of the previous curve (Fig. 4.2).

The large peak at 1280 Hz is generated by the wall to which the cement grout bulb is bonded, this response corresponds to a length of 80 cm.

4.2 Static Test and Bulb of Cement Grout

The following static test cannot permit to determine the internal tension force of the tie rod. The residual deformation is 3 mm and the unloading is linear. The operator indicated that the test had not lifted the head of the tie rod from the wall (Fig. 4.3).

J. J. Rincent, *Anchors—Tests Procedures and Vibratory Analysis*,
https://doi.org/10.1007/978-981-96-3777-5_4

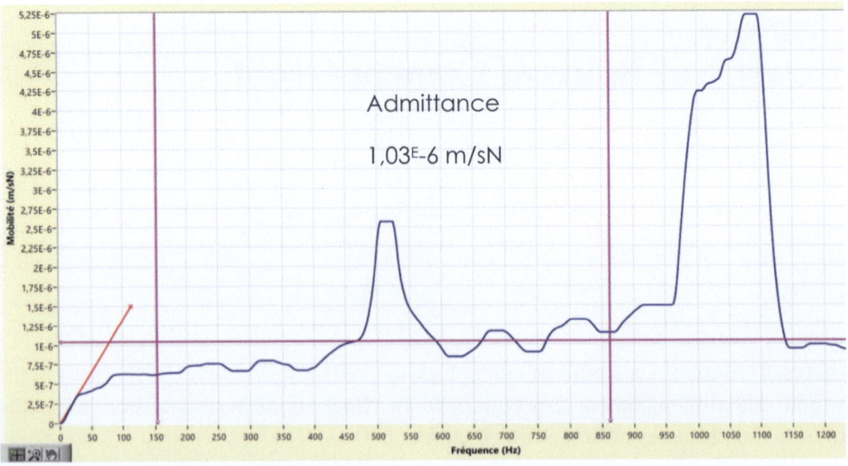

Fig. 4.1 Curve V/F function of frequency

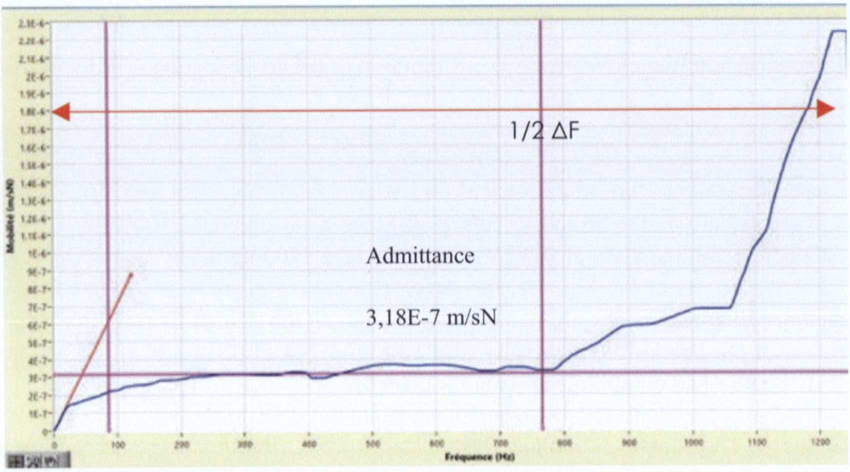

Fig. 4.2 Curve interpretation

The dynamic stiffnesses enable us to draw the curve below where the value of the internal tension can be determined. The tensile value appears during the loading phase, 15 tons (Appendix E) (Fig. 4.4).

It should be noted that the dynamic stiffnesses measured during simultaneous dynamic and static tests can often be used to define the value of the internal tension in the tie rod.

For the static test until 20 tons the grout bulb in contact with the wall and becomes deformed, a deformation permanent. The dischargement is linear and correspond to

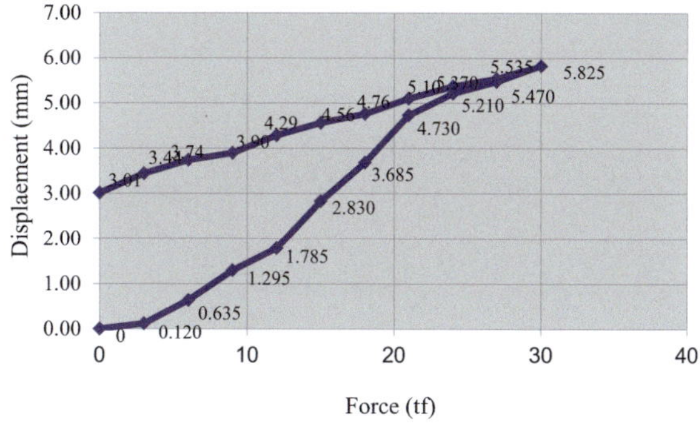

Fig. 4.3 Example of static test

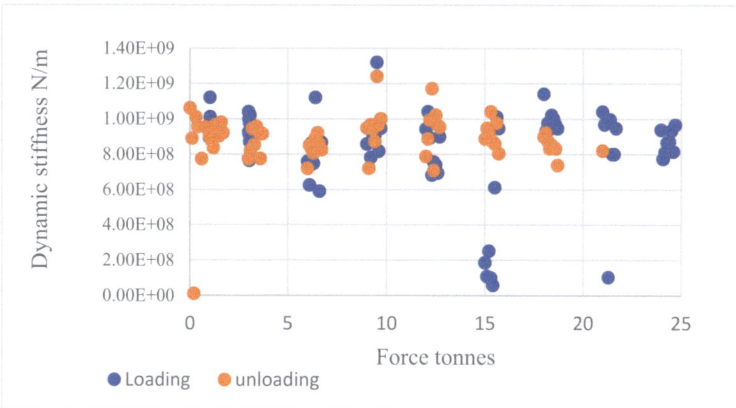

Fig. 4.4 T352 Dynamic stiffnesses N/m

the 32 mm bar of 1.6 m length which corresponds to the length utilized for the test plus the thickness of the wall and a part of the grout bulb length.

This tensile test demonstrated the existence of the cement grout bulb, caused the millimetre deformation of this bulb in compression on the wall and verified the elastic deformation of the bar, but did not give the internal tension of the tie rod (Fig. 4.5).

The presence of injection grout bulbs highlighted by the vibratory response curves is confirmed by drilling through the wall to the diameters calculated from the curves (Fig. 4.6).

On this site, 20% of the elements tested had large injection grout bulbs located against the retaining wall. In practice, this means that these tie bars cannot be re-tensioned, as the static tensile tests failed to detach the reinforcement from the grout, even with high forces.

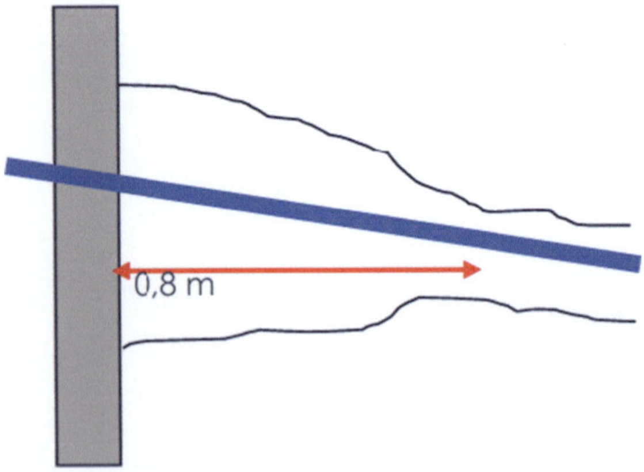

Fig. 4.5 Scheme of cement grout bulb

4.3 Others Examples

We have chosen another example where the static test does not allow us to determine the tension force internal to the tie-rod.

The curve below obtained on tie rod T15 does not allow the internal tension force of the tie rod to be defined during loading or unloading (Fig. 4.7).

Deformation 1 corresponds to the elastic elongation of the 32 mm diameter bar under tensile stress applied over a length of approximately 1.45 m, calculated from the dynamic test. Deformation 2 is the non-elastic deformations of the bulb located behind the wall where the tie rod is located (Fig. 4.8).

Points 1–8, tension force 1 ton; 9–16, tension force 8 tons.

During static and dynamic tensile tests, it should be noted that the values of 13–15 during the 8 tons stage are abnormally low. This phenomenon indicates that the tension force has been exceeded (Appendix F).

This procedure enables the value of the tension force to be identified or the value obtained from the static test to be confirmed. The many tests carried out lead us to prefer the value of the force to the discharge of the static test, because during loading, bonding forces lead to an overestimation of the tension force, a phenomenon stated when the tie rods are re-tensioned.

Monitoring and analysing the dynamic stiffnesses calculated during the loading test are a fairly accurate way of determining the internal tension force of the tie rod.

The following figure of the tension and the equivalent diameter of one part of a retaining wall shows:

- 19.6% of the tie rods have an equivalent diameter inferior at 0.60 m which is a high value
- 30.4% of the tie rods have a diameter higher than 0,9m which is abnormal
- 50% are between 0.6 and 1.2 m of diameter.

For example, for tie 47, the diameter of the tie with its cement grout is 0.5 m at the top of the first metre, where the equivalent diameter is 1.2 m (Fig. 4.9 and 4.10).

Fig. 4.6 Drilling confirm presence of the bulb. Source Rincent BTP Recife

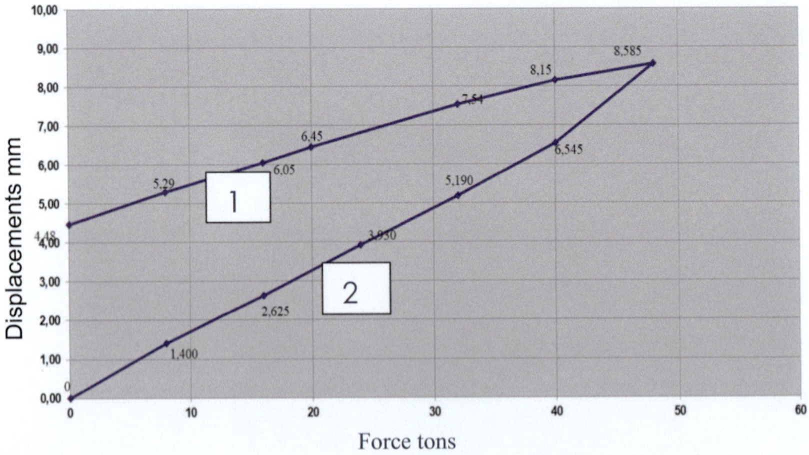

Fig. 4.7 T15 static test

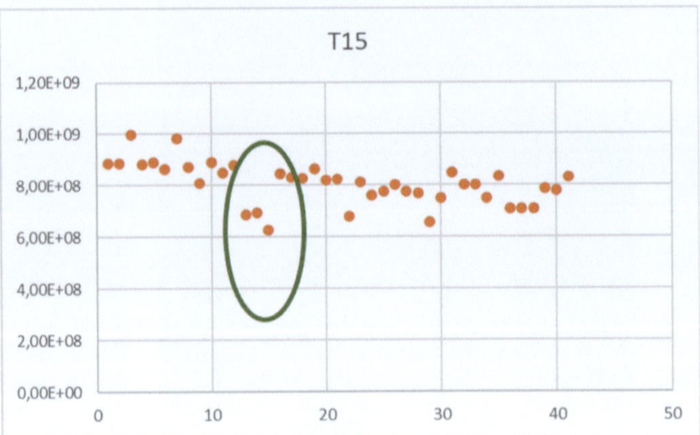

Fig. 4.8 Dynamic stiffness N/m

The consequences are:

- The presence of the bulbs means that the majority of the tie rods cannot be re-tensioned.
- The conventional static does not permit to determine the tension force of the tie rod
- The simultaneous tests static and dynamic with the stiffnesses values lead to obtain the internal force value (Fig. 4.11).

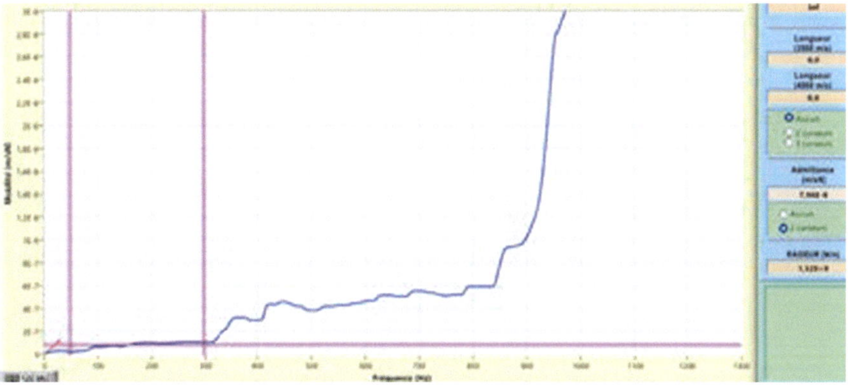

Fig. 4.9 First part of the curve: 7.96E−8 Equivalent diameter 1.2 m

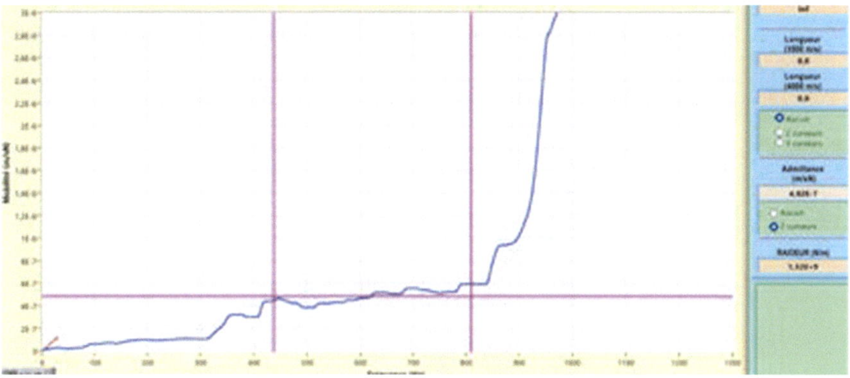

Fig. 4.10 Second part of the curve: 4.82E−7 Equivalent diameter 0.5 m

9,39	11,51	10,86	8,27	8,47	9,96	12,4	4,21	9,91
0,59	0,82	0,56	0,66	0,64	0,75	0,91	0,72	0,89
12,03	4,88	12,18	8,46	9,79	10,02	13,43	9,96	13,75
0,83	0,96	0,7	0,62	0,64	0,63	0,8	0,84	0,83
15,94	21,62	14,03	9,05	8,63	10,36	15,12	7,82	6,78
0,8	0,82	0,7	0,69	0,65	1,15	0,92	0,77	0,84
Tension force tons		9,6	13,06	8,63	12,97	12,73	9,73	14,71
Equivalent diameter m		1,23	0,94	0,64	0,91	0,76	0,90	0,86
		20,19	15,00	12,35	12,1	12,49	12,3	15,73
		0,67	0,60	0,91	0,81	0,8	0,93	1,02
				18,78	17,38	21,72	22,23	20,00
				0,97	0,8	0,83	0,93	0,77

Fig. 4.11 Tensions forces and calculated equivalent diameters

The histogram of equivalent diameters shows that most of these values are greater than 0.7 m, with an average of 0.8 m. This excessive volume of grout close to the wall has an impact on the static tensile test (Fig. 4.12).

The square roots values function of the equivalent diameters shown the influence of the bulb on the stiffnesses (Fig. 4.13).

In this location, a bulb with a diameter of 1 m and a length of 2 m generates an additional weight of about 3.5 tonnes compared to a tie rod with a diameter of 0.5 m.

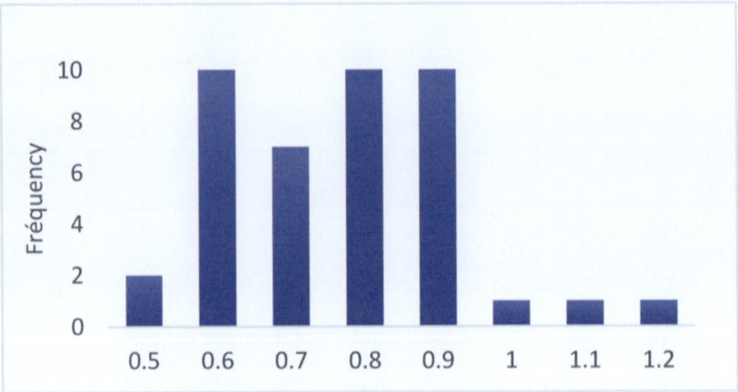

Fig. 4.12 Histogram of the equivalent diameters (m)

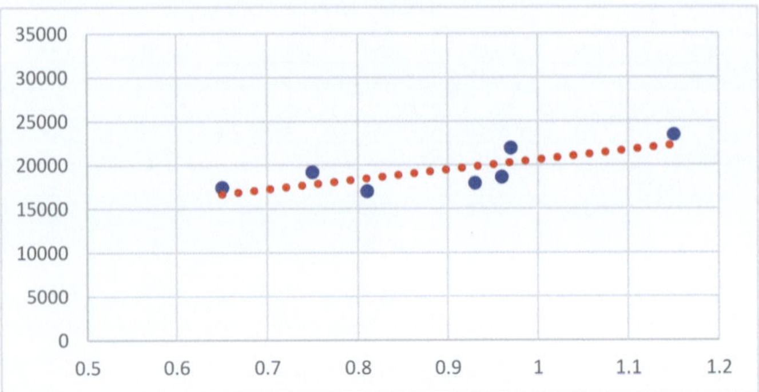

Fig. 4.13 Square root of stiffness as a function of bulb diameter

Chapter 5
Dynamic Stiffness/Static Stiffness

Abstract After a theoretical presentation of the dynamic and static stiffnesses, examples of the relationship between these two stiffnesses are given, as well as their use to determine the internal tension of the tie rod.

5.1 Dynamic Stiffness

The dynamic stiffness Rd is expressed in Newton per metre, i.e. the same units as a modulus, and is used to calculate the tensile forces in the tie rods (Fig. 5.1).

The dynamic stiffness calculated is a tangential stiffness, a complex number whose value is equal to $2\pi b/a$. A complex number is of the type $z = x + iy$, X the real part, Y the imaginary part. Studies carried out on the real and imaginary parts show that these values are constant for frequencies below 80 Hz. The range of frequencies used to calculate the dynamic stiffnesses of tie rods is less than 80 Hz and more generally less than 40 Hz.

Simultaneous static and non-destructive testing of a tie-rod produces the figure shown below, with the stiffness increasing as the element under test vibrates less (Figs. 5.2 and 5.3).

5.2 Static Stiffness

The static stiffness of the tie-rod under a given load is the value of the slope of the curve at the point corresponding to the load applied.

In this case, the static stiffness of tie 42 under a 20 tons load is 7.22E+08 N/m (Fig. 5.4).

J. J. Rincent, *Anchors—Tests Procedures and Vibratory Analysis*,
https://doi.org/10.1007/978-981-96-3777-5_5

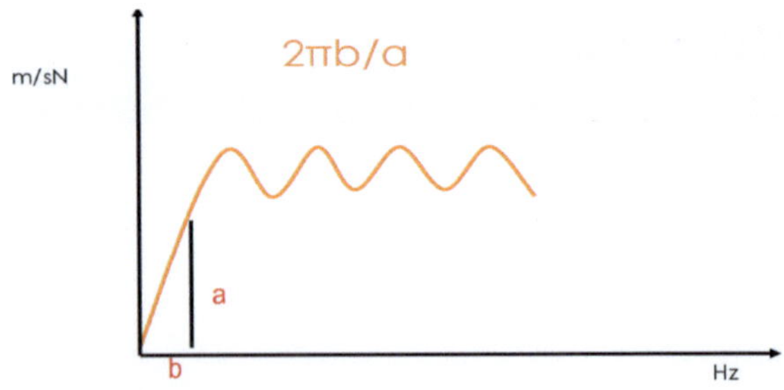

Fig. 5.1 Stiffness calculation

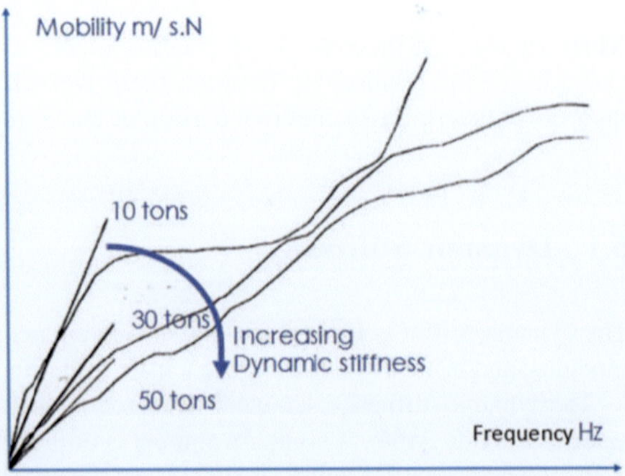

Fig. 5.2 Stiffness and traction force

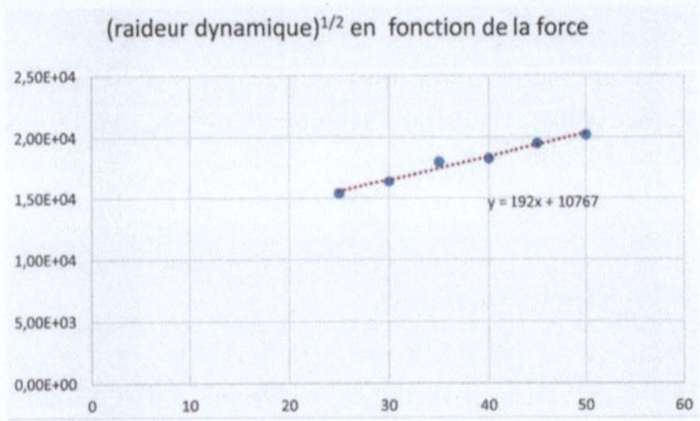

Fig. 5.3 Dynamic stiffness square root function of force

5.3 Dynamic and Static Tests

In order to carry out simultaneous static and dynamic tests, a metal structure needs to be designed which can:

- to be fixed to the head of the tie rod or by connecting the strands of the tie rod with strands that will ensure transmission of the force

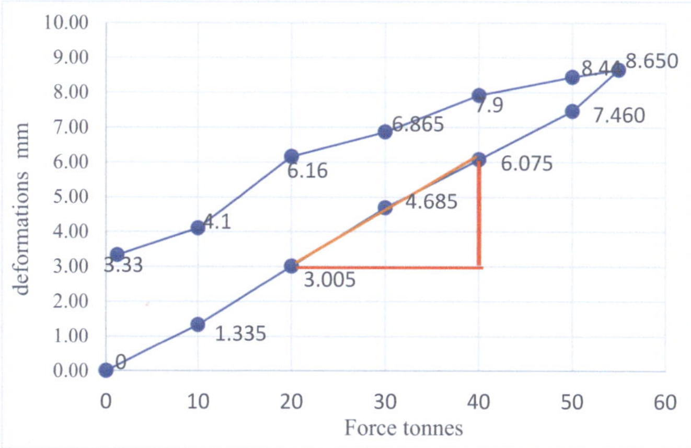

Fig. 5.4 Static stiffness

- while allowing non-destructive tests to be carried out at each level.

Tie rod 42 was first tested dynamically to obtain results under the existing initial site conditions. It was then tested simultaneously in static and dynamic conditions.

This procedure eliminates any disturbances that may be generated by the installation of the tensioning device.

The static tensile test was carried out as follows:

- Loading stages 1.2 t, 10 t, 20 t, 30 t, 40 t, 50 t, 55 t
- Unloading idem
- Duration of stops, loading 15 min
- Unloading 5 min.

It should be borne in mind that a tensile test carried out on a 35-year-old tie rod presents a risk and that the procedure to be implemented has nothing to do with, for example, a procedure for accepting a new tie rod for validation.

A detailed analysis of the evolution of dynamic stiffnesses at each load level leads to the graph below (Appendix G) (Fig. 5.5).

The simultaneous static and dynamic tests lead to the following comments:

- For example, to go from 20 to 30 tons, 8 hammer blows are made, causing a compression wave that induces physical changes to the system made up of the prestressed tie in its environment.
- The same applies from 30 to 40 tons, and so on.
- Each curve is analysed and used to calculate the dynamic stiffness.
- The tests are carried out over a period of more than 5 min and less than 10 min.
- On this site and for this dam, there are two families of points: hits 1 to 4 make up the first family and hits 5 to 8 the second family.

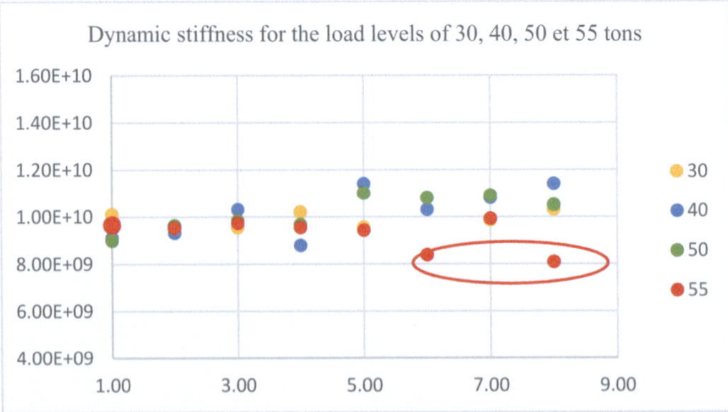

Fig. 5.5 Dynamic stiffness N/m for 30, 40, 50, 55 tons

- The variations and decrease in the dynamic stiffness of the tie rod when the tensile force is increased from 50 to 55 tons is characteristic of a change in behaviour as soon as the internal tension of the tie rod is exceeded.
- For the 55-tons test, two dynamic stiffness values are close to 8E9 N/m, whereas under the previous levels, the dynamic stiffnesses are all greater than 1E10 N/m.

As long as the tension force is not exceeded, the dynamic stiffness remains constant. If the internal force is exceeded, the way in which the tensile force is applied changes, causing a temporary drop in dynamic stiffness (Fig. 5.6).

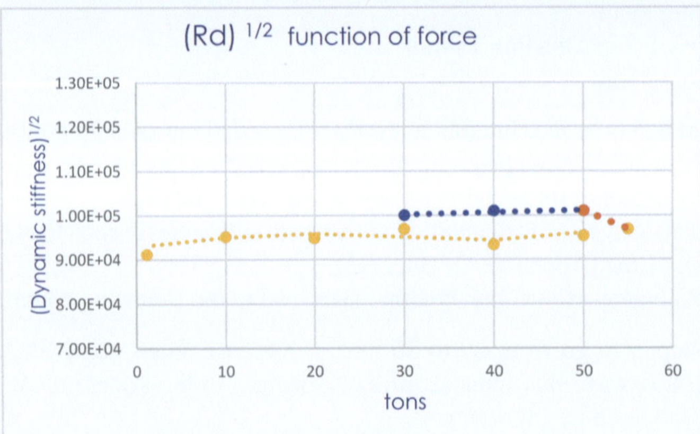

Fig. 5.6 Dynamic stiffness square root function of force

Fig. 5.7 Static and dynamic stiffness

	Dyn S N/m	Static S N/m	Ds/Ss
20t	8,79E+09	7,22E+08	12,2
30t	1,00E+10	7,22E+08	13,9
40t	1,02E+10	7,43E+08	13,7
50t	1,02E+10	7,91E+08	12,9

5.4 Dynamic Stiffness/Static Stiffness

The static stiffness is less than the dynamic stiffness and the ratio between the values of the dynamic and static stiffnesses is constant for the same element subjected to the same load in the pseudo-elastic behaviour phase.

Philippe Guillermain's 1972 thesis, carried out under the supervision of Allan Davis at CEBTP, shows that there is a direct link between dynamic stiffness and static stiffness: dynamic stiffness is greater than static stiffness, and there is a constant relationship between the two stiffnesses for an identical situation in the pseudo-elastic phase.

In the example of anchor 42, an anchor with 12 strands, the average ratio is 13 (Figs. 5.7, 5.8, 5.9 and 5.10).

Fig. 5.8 Ratio values

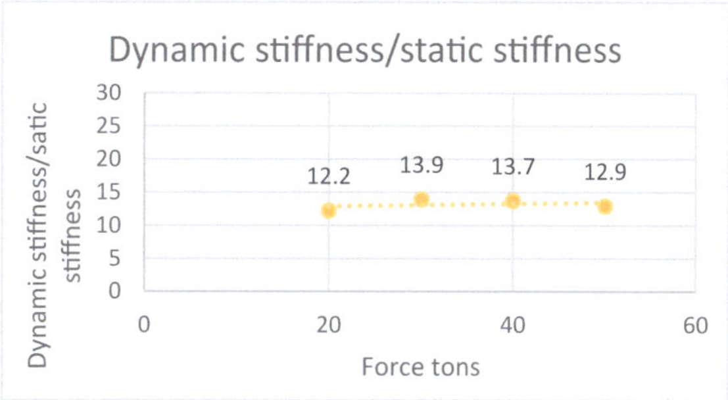

Fig. 5.9 Tie rod—rebar 32 mm diameter

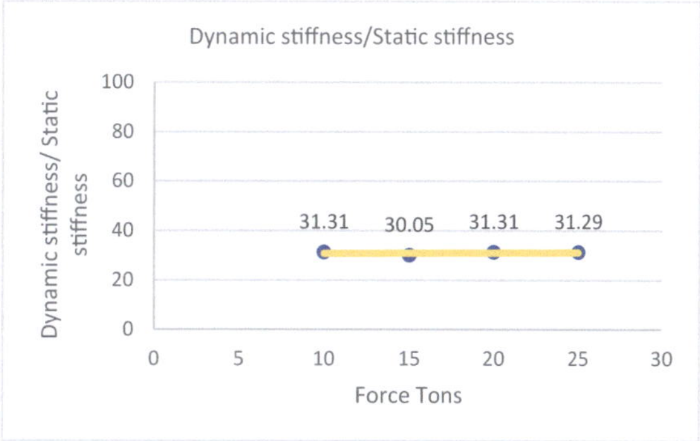

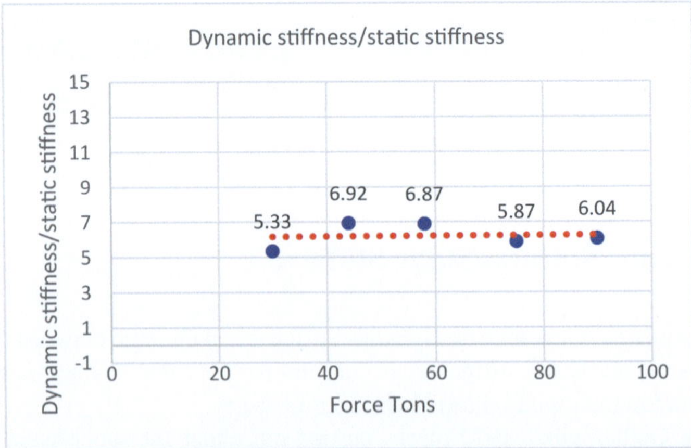

Fig. 5.10 Micro pile—petroleum tube109/127 mm

Chapter 6
Tension Force Calculation

Abstract A calculation of the internal tension of a tie rod is presented in a simple example. Other analyses show more complicated configurations. The example shown concerns horizontal tie rods running from concrete blocks to a sheet-pile curtain to explain the inertias combination.

6.1 Common Case

The aim is to calculate a and b in the formula $y = ax + b$, so two values of y and x are all you need to establish the formula linking the dynamic stiffness and the tension force of the tie-rod. Y is the square root of the tie rod dynamic stiffness and x is the tension force.

Tension force issued of static test and dynamic stiffness:

- T04 near of 5 tons
- T 73 near of 10 tons.

Dynamic stiffnesses:

- T04 2.07E8 N/m
- T73 5.38 E8 N/m

The formula obtained is $y = 1762x + 5579$.

On the same wall a third test has been performed on T 155. Tension force obtained from the static and dynamic tests is near of 10 tons.

The dynamic stiffness is 5.98E8 N/m, the calculated force is 10.7 tons seems correct.

The formula must be tested for example on the anchors with a null force, with the site information. The homogeneity of the backfill and the location of the tests in the central part of the wall are favourable factors in finding a suitable formula.

Meanwhile the site operator chooses the location tests with criteria favourable to direct tensile testing, for example the length of bar protruding from the head of the tie-rod.

J. J. Rincent, *Anchors—Tests Procedures and Vibratory Analysis*,
https://doi.org/10.1007/978-981-96-3777-5_6

6.2 Additional Analyses

We'll take the case of the concrete beams that exist on certain walls.

A pragmatic step consists of taking dynamic stiffness results for two tie rods, one with a beam and the other without. The two tie rods chosen are close in location and at similar heights (Figs. 6.1 and 6.2).

The square roots of the stiffnesses are: 38,079 for 245 with a beam and 24,228 for 265 without beam.

The difference of 13 850 is due to the additional inertia of the beam.

A series of tests on a beam element without tie rods was carried out (Fig. 6.3).

It is very important to know whether the beam is attached to the wall or not. The results of the first 7 impacts show that the answer is yes, the final impact shows that it is not. It is essential not to underestimate the influence of impacts on changes in the behaviour of structures in charge.

The dynamic stiffnesses of P1 are:

The average value of the 7 first stiffnesses is 2.2E8 N/m. The square root of the beam stiffness is 14,800, the result is fairly close to the previous value (Fig. 6.4).

The B value is linked to the wall, the anchor inertia and other element for example iron beam or concrete beam.

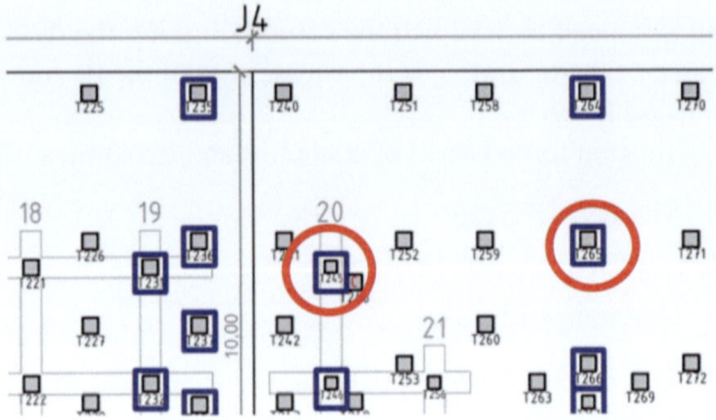

Fig. 6.1 Dynamic stiffnesses T245 and T265

Fig. 6.2 Dynamic stiffnesses T245 and T265

245	1,45E+09		265	5,81E+08	
	1,36E+09			6,14E+08	
	1,43E+09			5,84E+08	
	1,44E+09			5,75E+08	
	1,39E+09			5,67E+08	
	1,57E+09			5,74E+08	
	1,47E+09			6,15E+08	
	1,54E+09	1,45E+09		5,96E+08	5,87E+08

Fig. 6.3 Tests located on the concrete beam. *Source* Rincent BTP Recife

Fig. 6.4 T1 Dynamic stiffnesses results

2,51E+08
1,98E+08
1,06E+08
2,18E+08
2, 56E+08
2,91E+08
2,31E+08
1,29E+09

6.3 Inertia Combination

The example shown concerns horizontal tie rods running from concrete blocks to a sheet-pile curtain (Fig. 6.5).

The total stiffness measured in the axis of the horizontal tie-rod is the sum of the inertia of the mass and its contact with the soil in place and also the inertia of the tie-rod and the force existing in the tie-rod.

Each series of tests comprises 12 acquisitions, the results of which are averaged two by two for the analysis, which reduces random parasitic noise and amplifies the vibratory response of the element tested.

Two types of tests are carried out:

- A vertical test on the column, which characterises the column and its contact with the ground

Fig. 6.5 View of the concrete blocks. Rincent BTP France

- The other test was carried out horizontally on the same concrete block to characterise the concrete block and its contact with the ground, as well as the tie rod and the internal force.

The results are summarised in the graph below.

The stiffnesses of the anchors are very homogeneous, ranging from 4 to 4.5 10 8 N/m. The points in the graph correspond to stiffness measurements taken on two different dates, bearing in mind that at the second intervention, tie rods 2, 3, 5 and 6 were deactivated. It should be noted that tie rod 8 was already no longer working, since the dynamic stiffness measured during the first operation was very close to the value of the stiffness of the structure alone (Fig. 6.6).

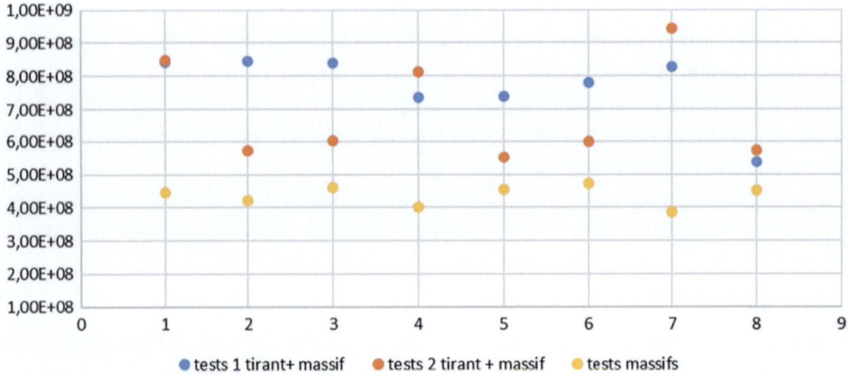

Fig. 6.6 Dynamic stiffnesses N/m Tests 1 et 2

Fig. 6.7 Dynamic stiffness square root

2	3	5	6
8,23E+03	7,45E+03	6,05E+03	6,15E+03
3,44E+03	3,10E+03	2,19E+03	2,73E+03
-58%	-59%	-64%	-56%

The stiffnesses are combined according to their square roots. Knowing the stiffnesses of the 'solid + tension rod' systems and the 'solid' stiffnesses enables us to identify the square root of the stiffness linked to the tension rod.

After deactivation, stiffness decreased by:

In this case, the part linked to the tension of the tie-rod represents approximately 60% of the total value of the stiffness (Fig. 6.7).

6.4 Curves Interpretation—Stiffnesses

In some tests, two dynamic stiffness values may appear in particular test phases, for example under low loads during simultaneous tests, either at the start of the test or during unloading in the final phase. The dynamic stiffnesses may correspond to the stiffness of the tie rod alone or of the tie rod plus that of the wall (Fig. 6.8).

Using the T 04 tie rod used previously, two dynamic stiffness values are identified. the internal tension of the latter is low and this type of configuration is often favourable to the calculation of two dynamic stiffness values, one corresponding to the stiffness of the tie bar surrounded by its injection grout, the other value incorporating the tension of the tie bar.

The formula adopted for this group of tie rods is:

$$y = 2260x + 2848$$

which leads to an internal force of 5.11 tons, confirmed by the static test (Fig. 6.9).

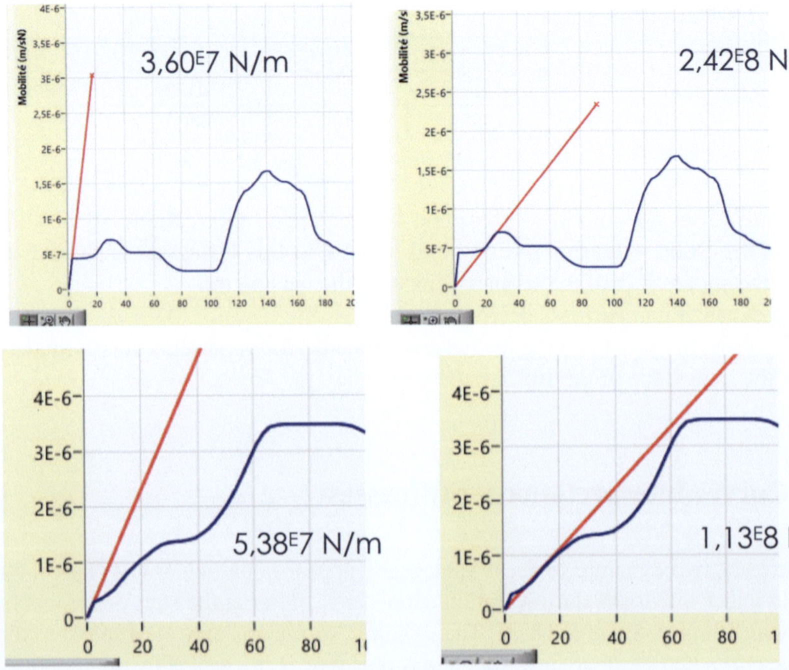

Fig. 6.8 Curves examples

Fig. 6.9 T04 Dynamic stiffnesses

T04	9,09E+06		2,25E+08		
	1,68E+07		2,02E+08		
	8,10E+06		2,54E+08		
	7,01E+06		2,01E+08		
	1,05E+07		2,00E+08		
	6,91E+06		2,07E+08		
	5,71E+06		2,06E+08		
	7,04E+06	8,11E+06	2,05E+08	2,08E+08	5,11t
		2848			

Chapter 7
Curves Response—Analysis

Abstract The vibration response led to analyse the behaviour of the anchor, several examples are given, the relative uncertainly of the dynamic stiffness is a parameter. The analyse of the response during the static tests is also used to follow the change of behaviour.

7.1 Dynamic Stiffness—Relative Uncertainty

On a 40-year-old retaining wall, details of the dynamic stiffnesses of the central part of the wall can be used to analyse the behaviour of the tie rods under these dynamic loads. The table below shows the values of the internal tension forces and the relative uncertainty calculated for 6 values of the dynamic stiffnesses i.e. with the highest and lowest values removed (Appendix H) (Fig. 7.1).

If the 8 stiffnesses values are maintained for the relative uncertainly the results for the line 6 are:

8 values	29,46%	60,87%	11,72%	7,75%
6 values	18,68%	41,28%	8,79%	6,97%

Analysis of the stiffnesses of lines 7 and 8 of tie rods enables us to calculate the relative uncertainty of four anchors with an internal force exceeding 20 tonnes.

Those values are below 15% and qualify the homogeneity of the measured stiffnesses. Tie rods with high tensions and low sensitivity to hammer impacts are tie rods that can potentially be re-tensioned.

For the ground anchors with a tension force between 10 and 20 tons the sensibility to the dynamic test is highly variable. More than 60% of anchors have relative uncertainties of less than 15% and are therefore stable under dynamic loading.

Let's look at the two tie rods on lines 6 and 8 with relative uncertainties on the dynamic stiffnesses of 81.43% and 41.28% respectively (Fig. 7.2).

J. J. Rincent, *Anchors—Tests Procedures and Vibratory Analysis*,
https://doi.org/10.1007/978-981-96-3777-5_7

line				
1	7,2	7,7	3,8	1,6
	8,91%	10,14%	31,94%	25,83%
2	10,4	10,1	8,6	11,3
	18,71%	5,66%	5,84%	6,19%
3	3,5	8,7	1,6	10,8
	35,30%	5,39%	21,90%	6,96%
4	4,6	0	1,1	11,5
	38,06%	17,66%	30,64%	15,48%
5	10,9	11,3	10,2	14
	11,33%	3,93%	27,57%	5,74%
6	13,9	12,9	12,2	12,2
	18,68%	41,28%	8,79%	6,97%
7	18,7	23,2	23,6	18,4
	21,60%	13,90%	7,49%	3,93%
8	26,3	13,4	20,5	23
	11,88%	81,43%	10,19%	6,95%

Fig. 7.1 Result table—forces (tons) relative uncertainly %

In both cases, the first three strikes of the hammer cause a reduction in the dynamic stiffness, which then increases again. It is possible to put forward the hypothesis of the destruction of lateral friction, which is then remobilised.

7.2 Vibration Response During the Static Test

The example of the static test carried out on tie rod T49 (Appendix I) shows the contribution of the analysis of the vibratory response and the parameters calculated from these curves.

The internal tension force of the tie rod cannot be determined from the curve obtained from the static test (Fig. 7.3).

The dynamic stiffnesses are calculated for each hammer blow:

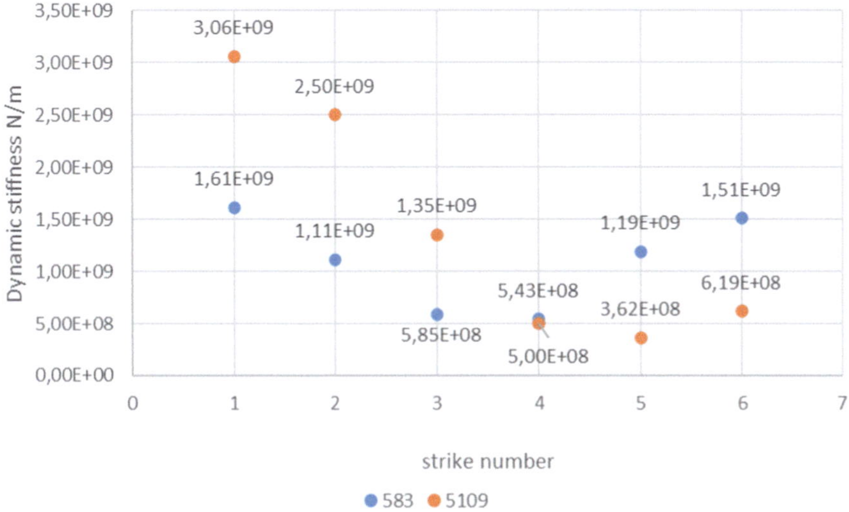

Fig. 7.2 T 583 and 5109 Dynamic stiffnesses

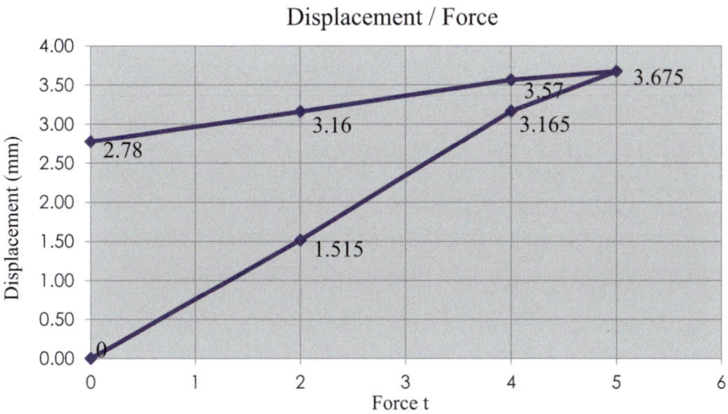

Fig. 7.3 Displacement function of force

- 1–8 initial tests
- 9–16 tension force 1 ton
- 17–24 tension force 2 tons
- 25–31 tension force 4 tons.

This dynamic stiffness curve combines the initial tests and then the tests under different tensile forces. A first analysis of the evolution of the stiffnesses, which begin to decrease during the one-tonne stage, shows that the internal tension force is less than 1 tonne (Fig. 7.4).

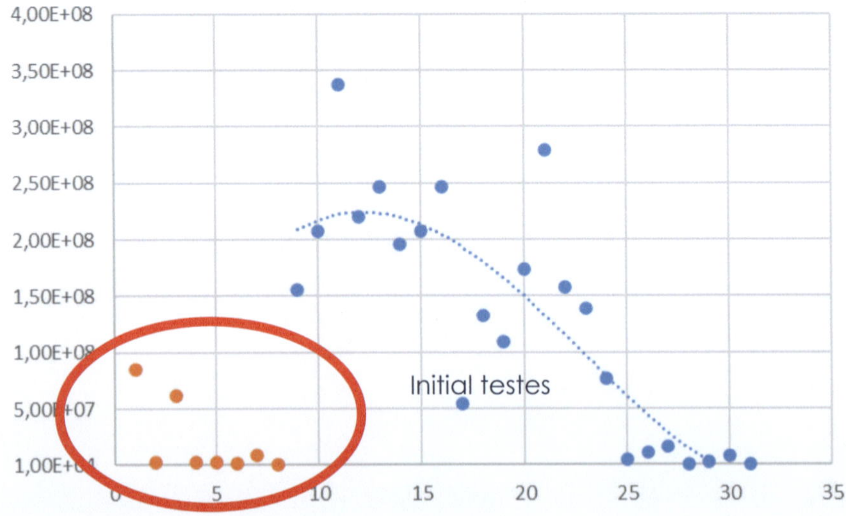

Fig. 7.4 Dynamic stiffness N/m under different forces

Analysis of the curves shows the evolution first curve shows the tie rod total length of 20 m and the response of the wall with a thickness (Figs. 7.5, 7.6 and 7.7).

A partial rupture of the skin friction increases in the part of the free length from 8.7 to 9.7 m.

At the first blow at the step of 1 ton, the mechanical admittance is 2.21E−6 m/sN which corresponds to an equivalent draught diameter of 0.24 m. On the curve obtained after the rupture, the free part still appears but reaches 12.8 m in length, the total length is unchanged but the mechanical admittance is 2.6E−4m/sN. This is

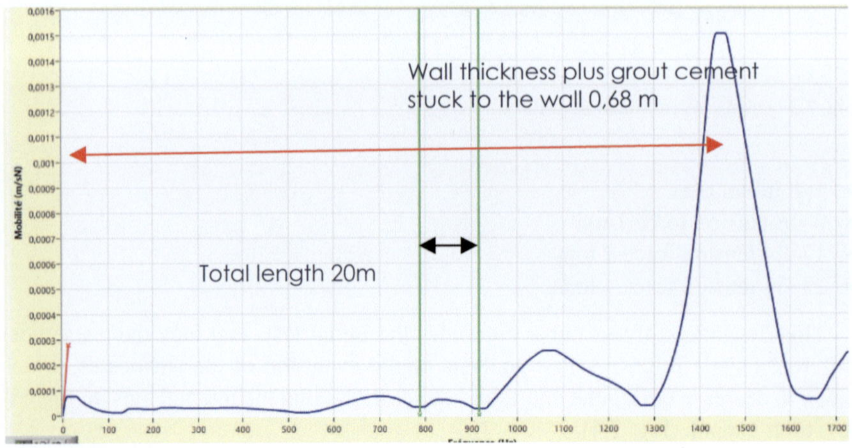

Fig. 7.5 Initial test

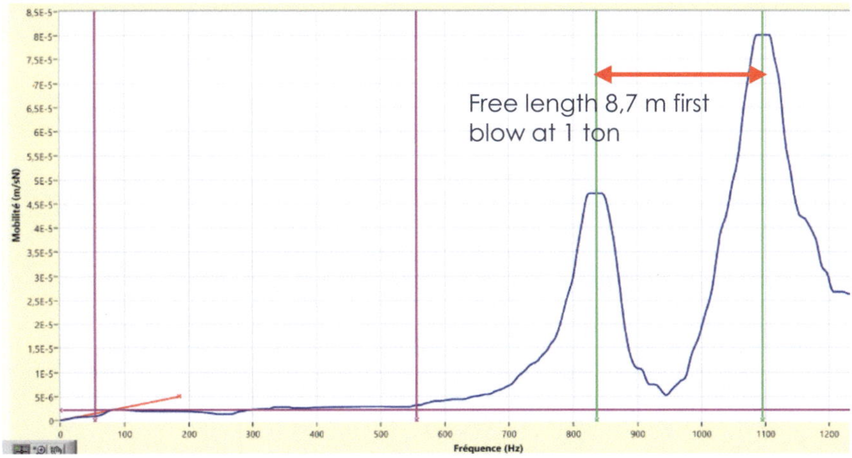

Fig. 7.6 Tie rod free length

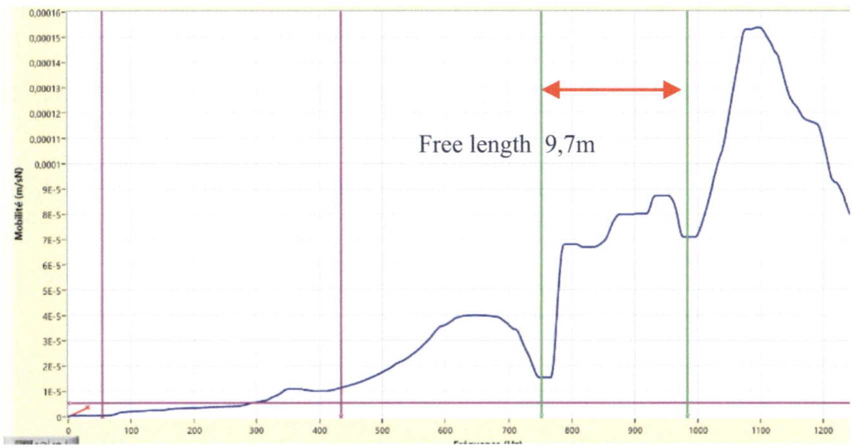

Fig. 7.7 Partial rupture of the skin friction

the diameter of the bar detached from the cement grout, i.e. 3 at 4 cm of diameter in function of hypothesis taken (Fig. 7.8).

The complete analysis of the curves shows the physical changes induced at each stage, whether loading or unloading.

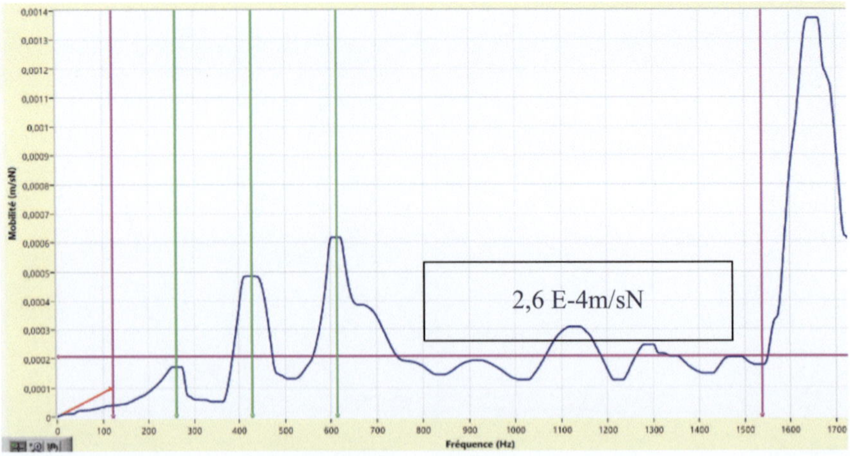

Fig. 7.8 After rupture, length and mechanical admittance

7.3 Static Test—Unloading

When we first started providing services more than 30 years ago, we only carried out simultaneous tests during loading and not during unloading. We found that we were losing information that was useful for understanding the test. For example, for the Tie rod 155 at unloading, the determination of the internal force is clearly 10 tonnes. This point cannot be identified during loading, including during the static test (Figs. 7.9 and 7.10).

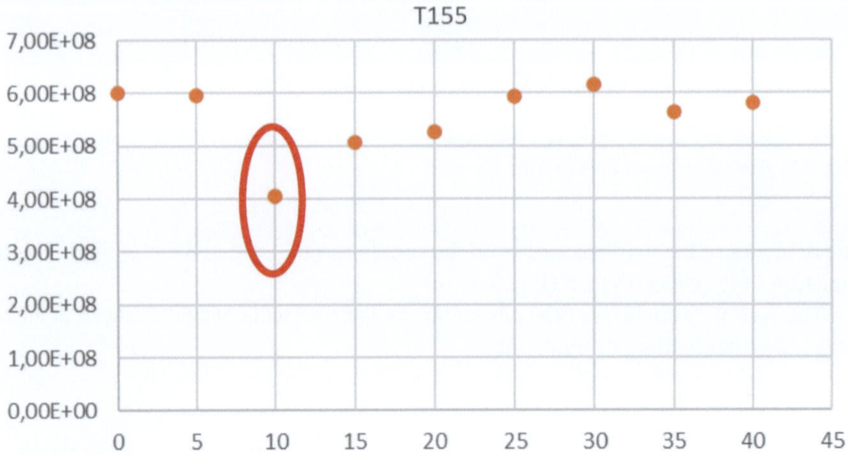

Fig. 7.9 Stiffness function of the load

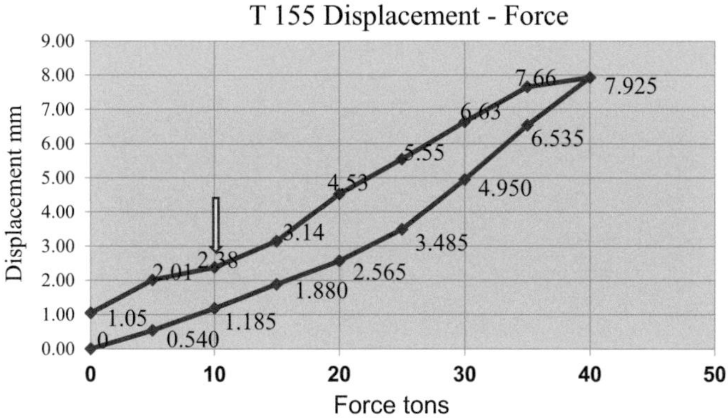

Fig. 7.10 Displacement function of force

Chapter 8
Uses of the Static Tests

Abstract The static tests are used to determine the tension force and also to know the ground anchor deformations. The deformation values will be used when re-tensioning. In addition, the stability of the tie rod at each level is an indication of the future behaviour of the ground anchor. Several site results are given.

8.1 Before Static Tests

Before performing static test, it is necessary to estimate the tension force which will be meet.

Generally, the total length and the free length are known from the dynamic tests the diameter also. A hypothesis on the skin friction leads to calculate the force which can be mobilised.

This force is used to define the means to be used, both in terms of equipment and, above all, the connection to be made on the tie rod to carry out the tensile test. In the case of a bar that can be extended by means of a sleeve, this is easy to do if the connection is long enough.

On the anchors whose reinforcement is made up of strands, the connection is more difficult, especially as the forces are often higher.

8.2 Results Analyses

The objectives of the static tests are:

- firstly, to determine the tension forces in the ground anchors
- to calibrate the non-destructive test method
- to obtain results on the deformations obtained under different loads and to derive behavioural laws for re-tensioning
- carry out these tensile tests at higher values for tie rods likely to be re-tensioned.

J. J. Rincent, *Anchors—Tests Procedures and Vibratory Analysis*,
https://doi.org/10.1007/978-981-96-3777-5_8

These tensile tests were not carried out to determine the internal tension force, but mainly to understand the behaviour of the tie rods under the loads applied. Here one simple cycle of loading was applied. Remember that these are site tests (Fig. 8.1).

The purpose of the site tests carried out on these 5 tie rods was to determine their behaviour under different loads, and cannot be assimilated to creep tests. We used the deformations measured between 10 and 15 min of loading under 4, 8, 12, 16, 20 and 25 tons. The following figure shows the stability under load of a part of the anchors (Fig. 8.2).

Tie rod	Strands	Force max	deformation	residual def.
number		tons	mm	mm
238	8	20	15,8	10,3
3127	8	20	24,8	10,35
3132	10	20	27,9	15,2
5125	10	25	15,1	4,4
5138	10	25	15,8	8,57

Fig. 8.1 Table of results

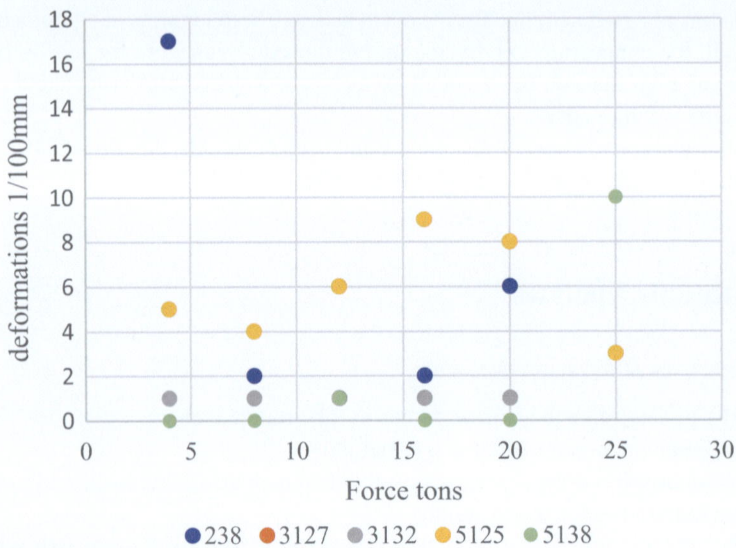

Fig. 8.2 Deformations function of the traction force

N° Cycle	Strands	Force tons	deformation max mm	deformation residual
138 1	10	11,5	5,2	2,3
138 2		11,8	4,1	0,77
261 1	8	15	14,4	4,4
261 2		15	7,7	0,4

Fig. 8.3 Maximum and residual deformation

The anchor stability can be measured with the relative uncertainty (Appendix J))

- 238 6.89%
  3127 20.30%
- 3132 3.40%
- 5125 4.74%
- 5138 11.77%

Here two loading cycles are applied (Fig. 8.3).

8.3 Deformation Rules

The aim was to test these tie rods and put them back under tension. These tensile tests were used to establish the following curves, one for 8-strand tie rods and another for 10-strand tie rods (Figs. 8.4 and 8.5).

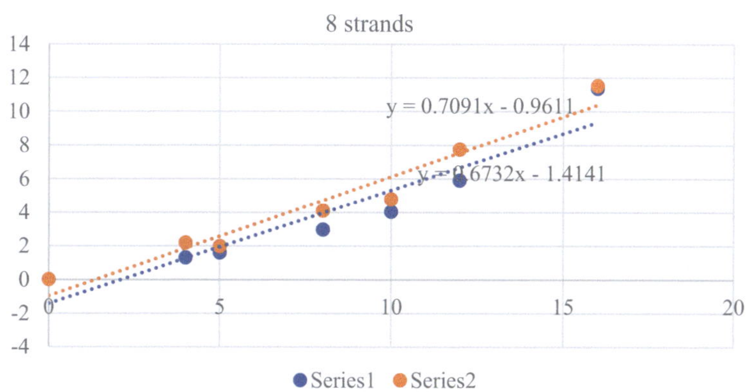

Fig. 8.4 Anchors 8 strands

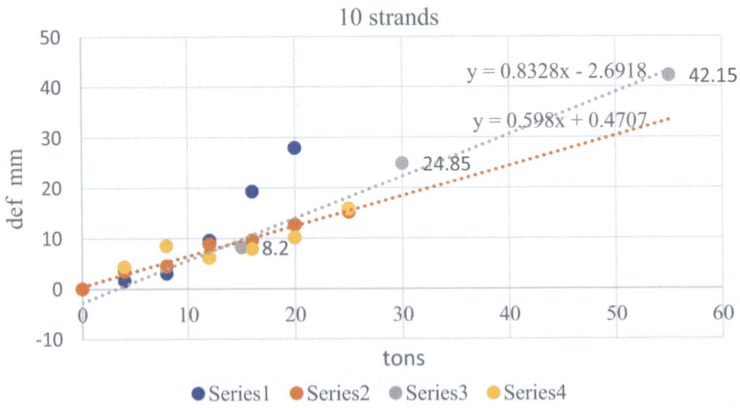

Fig. 8.5 Anchors 10 strands

Based on the stability of the dynamic stiffnesses, certain tie rods were identified as being suitable for re-tensioning. Cyclic static tests were carried out. Creep analyses were done, bearing in mind that the stages were shortened for reasons related to work at height, which limits the duration of the interventions (Fig. 8.6).

Analysis of the behaviour of the ground anchors during the tests is essential to define the tie rods that can support this additional load lost over the years. The relative uncertainty of the dynamic stiffnesses is the first element, and the results of the static tests are the second point of information.

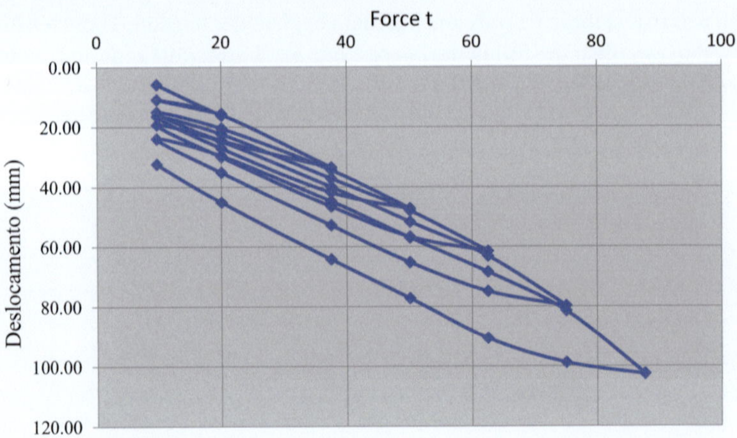

Fig. 8.6 Cyclic loading test

8.4 Equipment to Perform the Tests

An important point that will not be developed here is the design of the metal structure required to carry out the tensile test. As mentioned previously, this equipment must be relatively easy to handle in order to carry out tests at height. It must allow the head of the tie rod to be tensioned by resting on the wall.

Finally, the method of connecting the head of the tie bar must be adapted on a case-by-case basis for tensioning. In general, the length of the strands or bars is insufficient. Very often the choice of anchors to be tested in tension is linked to favourable configurations. For example, when the length of the part of the bar outside the head of the tie allows a fitting to be screwed on or other opportunities.

The future design of the tie-heads must allow easy access for testing and re-tensioning. A removable guard covering the tie-rod armature and protecting it with anti-corrosion grease for example and enabling dynamic tests to be carried out (Fig. 8.7).

Fig. 8.7 Schematic diagram

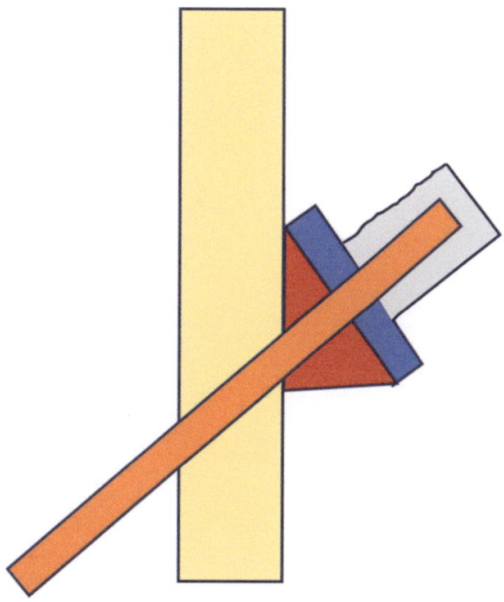

Chapter 9
Repro-tension

Abstract After explaining the use of electrical insulation measurement, the technical and economic benefits of re-tension are presented. An example of re-tension is given, describing the means used and the procedure.

9.1 Electrical Isolations

There is a Swiss standard for the acceptance of recently built tie rods with regard to their protection against corrosion. We have used this standard to identify tie rods that can be re-tensioned and those that cannot be re-tensioned because they do not comply with the standard. The standard gives results guidelines and there is a test standard that specifies the procedure and means to be used.

Extract from the revised version of the Geotechnical SIA 267 has been in force since 1 August 2013.

10.7.4.1 Tie rods fitted with corrosion protection in accordance with figure 10.6.3.2 shall be subject to a check of the effectiveness of the corrosion protection Electrical Resistance Measurement I (ERM I), after injection and connection. The check shall be carried out in accordance with standard SIA 267/1.

10.7.4.2 A tie rod, once injected and connected to voltage, must have an electrical resistance RI 0.1 MΩ (Mega Ohm).

10.7.4.3 In the event that a certain percentage of tie rods do not have the electrical resistance I is tolerated, for example 5–10%, an electrical resistance measurement II (ERM II) will be carried out on the defective tie rods to prove that there is no galvanic contact between the head of the tie rod and the frames of the structure. The check will be carried out in accordance with standard SIA 267/1.

The electrical resistance between the tie rod head and the structural frames must have a value of RII 100 Ω.

10.7.4.5 Corrosion protection requirements for tie rods with corrosion protection Limited corrosion in accordance with 10.6.3.3 will be checked visually before, during and after the placement of the tie rods.

Our proposal was to check the first requirement, which is to measure the electrical resistance of the tie rod.

J. J. Rincent, *Anchors—Tests Procedures and Vibratory Analysis*,
https://doi.org/10.1007/978-981-96-3777-5_9

We carried out a limited number of tests and found that a large number of tie rods complied with this standard, even after several decades in existence.

9.2 Interest of Repro Tension

The many tests carried out show that the number of tie rods where it is possible to add additional tension is very large. Firstly, these correctly insulated tie rods withstand the re-tensioning.

Current calculation methods make it possible to optimise the value of the additional tension to be applied. On one of the sites treated, it appears that the stability coefficient of certain walls is currently 1.05. Adding a force to achieve 85% of the initial tension results in a coefficient of safety of 1.55.

This operation costs 10–20 times less than making new tie rods. It is quite possible to carry out a combination of re-tensioning and new tie rods. The aim is to ensure the longevity and maintenance of the structures by increasing their durability.

It is also useful to monitor the loss of tension in these tie rods by carrying out dynamic tests planned over time, in order to carry out maintenance operations.

9.3 Procedure

The value of the force to be reintroduced into the tie rod is given by the design office, the internal tension of the tie rod is obtained from the non-destructive tests, so the difference in force is known. The additional deformation of the tie rod can be known from static tensile tests and the thickness of the metal plate to be inserted is defined.

The next step is to design and manufacture a metal structure that allows the head of the tie-rod to be pulled and the additional tensioning plate to be inserted (Figs. 9.1 and 9.2).

9.4 Plate Thickness Calculation

The example chosen concerns a 10-strand tension cable, the link between the deformation and the tensile stress can be formalised as follows:

$$y = 0.8x - 1 \quad \text{y deformation mm} - \text{x force tons}$$

Tests see previous paragraph.

The selected anchor has an internal tension force of 12 tons, for example the force to re-introduce is 33.6 tons i.e. an additional force of 21.6 tons which will produce an additional deformation of 16.3 mm. This calculation leads to choose plates thicknesses to introduce between the wall and the anchor head (Fig. 9.3).

Fig. 9.1 Tensioning device. *Source* Rincent BTP Recife

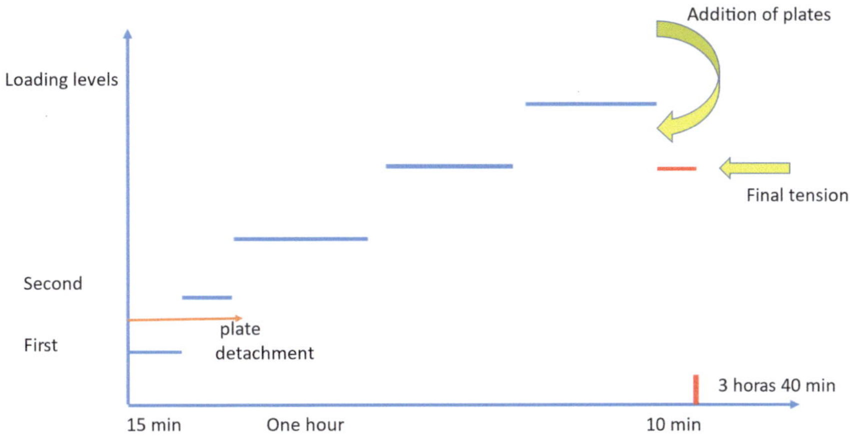

Fig. 9.2 Re-tensioning procedure

9.5 Particular Cases

The tension force of anchors 450 and 451 are:

	ND tests	Static tests
450	1.3t	9.0t
451	0t	2t

The static tests and the general shape of the curves show that these tie rods have not been tensioned or have been relaxed very quickly after their initial tensioning and that they can withstand high tensions (Fig. 9.4).

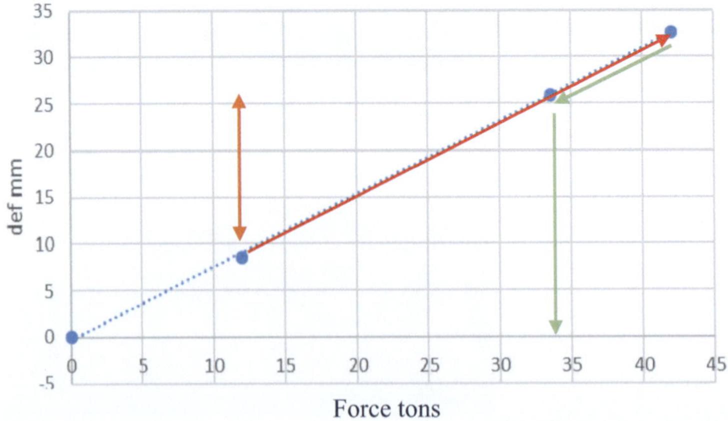

Fig. 9.3 Tension procedure

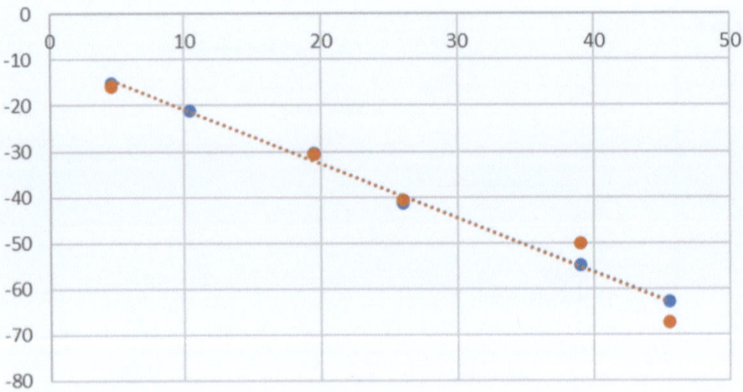

Fig. 9.4 450 and 451 deformations (mm) function of tensile force (ton)

9.6 Durability

Experience gained on construction sites in Brazil shows that tie rods 35 and 40 years old behave in a way that allows us to expect a longer service life than is currently envisaged.

In none of the cases studied in the last 5 years in Brazil was the quality of the concrete in the walls found to be non-conforming.

Re tensioning tests have been successfully carried out and the behaviour of the tie rods over time needs to be monitored. It should be remembered that the cost of re-tensioning tie rods is in no way comparable to the cost of installing new tie rods.

Fig. 9.5 Passage of a convoy. *Source* Rincent BTP Recife

Monitoring based on vibration analysis tests can be used to prevent any problems and plan maintenance of the structures.

At one of our test sites, a stability calculation of the retaining walls shows that these structures are very close to equilibrium, i.e. with a safety coefficient very close to 1. A re-tensioning of 15% less than the initial tension would allow a return to a safety coefficient greater than 1.5.

This is the result of more effective diagnostic and calculation methods that are closer to the reality of the site.

During load tests, we noted the impact of rolling loads on the measurements taken. The load of the wagon axles induces an effort close to 1 tonne, i.e. not very high. 10 convoys of 200 wagons with two axles for 300 days and 40 years generates 4.8E7 loads. The assumptions can be modified, the fact remains that the number of cyclical charges is high given the number of years of activity (Fig. 9.5).

On two sites, we have observed and measured the effects of rolling loads on the tension of tie rods on both road and rail traffic. The fatigue phenomenon needs to be studied in greater depth. The tie rods of structures subjected to cyclic loads need to be monitored more closely using instruments, with stresses being counted and tie rod tension monitored more frequently, etc.

This is how the maintenance of structures with tie rods will be managed in a sustainable way.

Several conclusions can be drawn from these tests:

- The electrical insulation of a ground anchor, even after several decades, can comply with a standard drawn up for new tie rods,
- On this site, the tension force resulting from the dynamic tests was slightly lower than the value measured during re-tensioning, at an average of 0.8 tonnes,
- And the values obtained from the static tests were slightly higher, averaging 1.2 tonnes.

Chapter 10
Pre-tensioned Dowels

Abstract The method used for ground anchors was originally used for pre-stressed fixings for industrial robots, and later for fixing the feet of pylons. Examples are given to show the possible uses.

10.1 Laboratory Tests

We carried out laboratory tests on prestressed dowels embedded in a concrete mass. The aim was to find out whether variations in dynamic stiffness could herald failure and therefore prevent it. The test set-up was as shown in Figs. 10.1 and 10.2.

Measurements of displacements per stage have enabled force–deformation curves and creep curves to be drawn.

The aim of these tests was to compare static and dynamic tests, and in particular creep curves and dynamic stiffness curves as a function of tensile force.

In this example, the static test shows that creep begins at 1600 kg and the creep load is 1800 kg (Figs. 10.3 and 10.4).

The results of these tests show that the dynamic stiffness measured at the same time as the static test provides an indication of the variation in stiffness, which is a precursor of a change in the behaviour of the anchor.

These results need to be studied in greater depth, because in practice, this approach would guarantee that the element under test would not break. For example, a cladding plate fixed to a wall could be tested without being damaged.

10.2 Pre-tensioned Bolts

Measurements of the dynamic stiffnesses of prestressed dowels can be used to identify connection problems. For example, on a turbo alternator fixed by bars passing through a 1-m-thick raft, vibration problems were identified at start-up and shutdown.

J. J. Rincent, *Anchors—Tests Procedures and Vibratory Analysis*,
https://doi.org/10.1007/978-981-96-3777-5_10

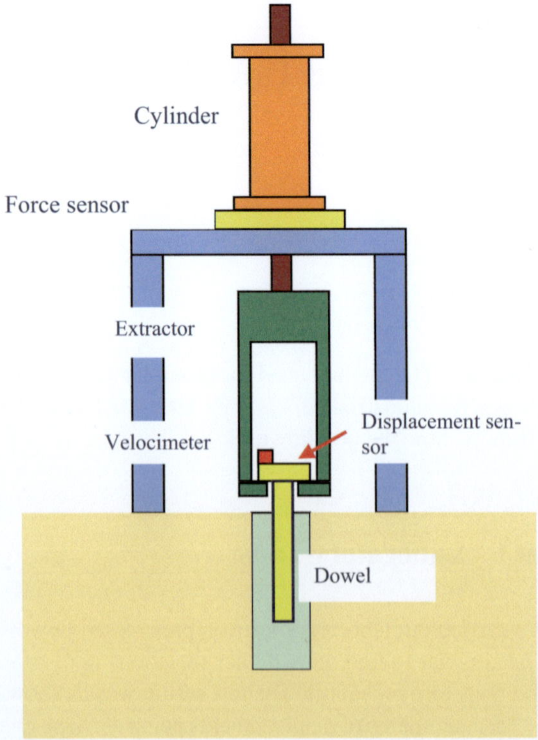

Fig. 10.1 Test scheme

The bar tensions were compliant, but the dynamic stiffness measurements detected significant disparities (Fig. 10.5).

Several litres of resin were injected at the periphery of these bars to re-establish contact with the concrete raft, and the stiffnesses became higher and more homogeneous.

1. The tension force is correct but the stiffness is low, the annular space is empty,
2. The tension force is correct and the stiffness is high, the annular space is filled with resin (Figs. 10.6 and 10.7).

These laboratory tests enabled us to familiarise ourselves with in situ tests, whether for road signalling pylons or pylons supporting radar, for example. The procedure is the same as for the tie rods, the means used can be identical to those used for the tie rods. However, a few adjustments of the materials used can be done.

A study of the influence of cyclic loads on the dynamic stiffness of the prestressed dowels has been started but has not yet been completed.

Fig. 10.2 Test devices. *Source* Rincent BTP France

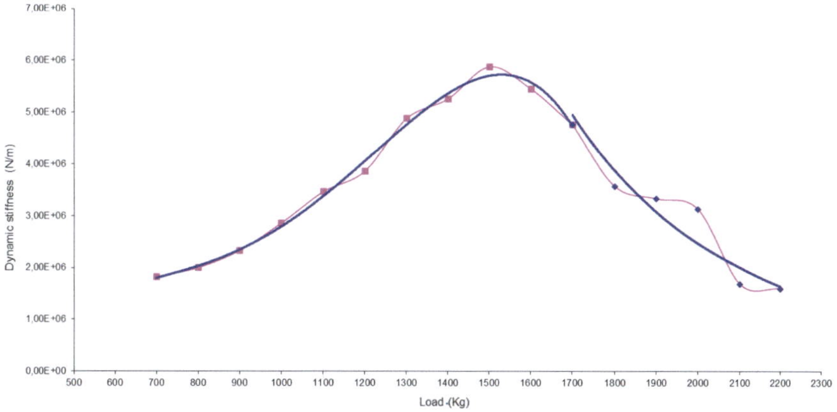

Fig. 10.3 Dynamic stiffness function of force

10.3 Pylons Fixations

The hammer fitted with a piezoelectric sensor is smaller in size for striking the head of a nut and the velocity sensor is also smaller, in this case it is often an accelerometer of small dimension (Figs. 10.8, 10.9, 10.10 and 10.11).

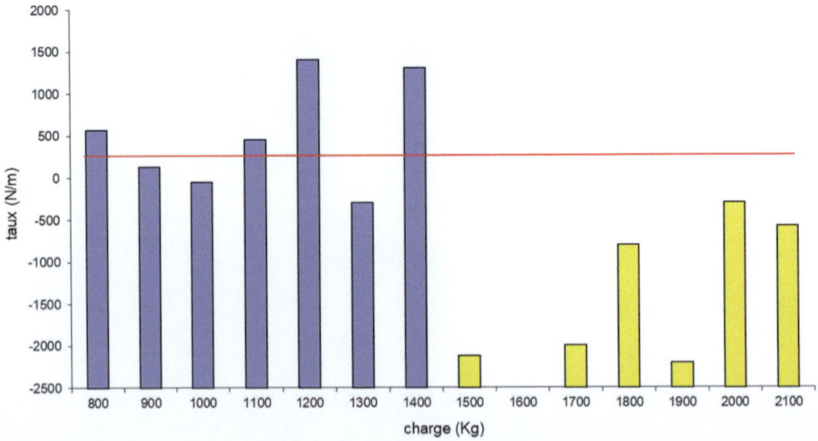

Fig. 10.4 Stiffness variation function of the force

Fig. 10.5 Dynamic stiffness table

n°	N/m
1	8,88E+08
2	1,03E+08
3	1,15E+09
4	1,45E+09
5	1,02E+09
6	7,84E+08
7	8,64E+08
8	8,74E+08
9	9,11E+08
10	1,05E+09

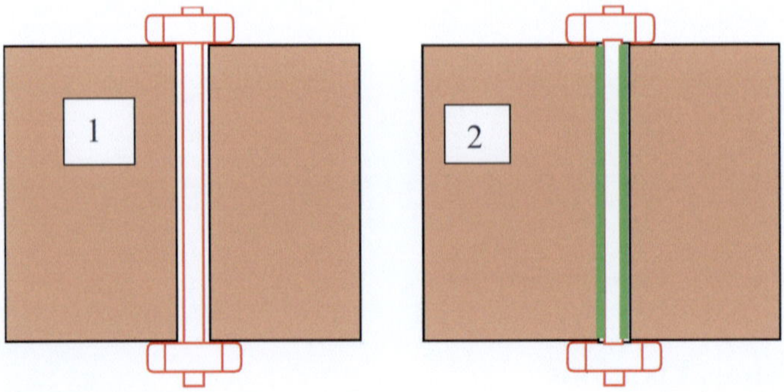

Fig. 10.6 Slab scheme

Fig. 10.7 Increased dynamic stiffness

n°	N/m	Increase %
1	1,00E+07	6%
2	1,33E+09	29%
3	1,52E+09	32%
4	1,54E+09	6%
5	1,18E+09	16%
6	1,04E+09	32%
7	1,27E+09	47%
8	1,32E+09	51%
9	2,01E+09	121%
10	1,37E+09	30%

Fig. 10.8 Road signalling gantry. *Source* Rincent BTP France

Bolts n° 3, 4, 19 and 32 with stiffnesses slightly below the threshold value (average-standard deviation) these bolts must be tightened, including those below average. An annual test campaign can provide information on changes in the tension of the anchor bolts and their position in relation to prevailing winds, for example. this radar tower is located on the North Atlantic coast (Appendix K).

10.4 Floor Mounting of Industrial Robots

Very early as 2003, we carried out tests on robot foot fixings in the concrete floors of several European car factories. The anchors are pre-stressed to minimise the movements induced by the robot arm when, for example, it positions a windscreen on the metal frame in order to bond it.

Fig. 10.9 Radar tower

Fig. 10.10 Instrumented hammer. *Source* Rincent BTP France

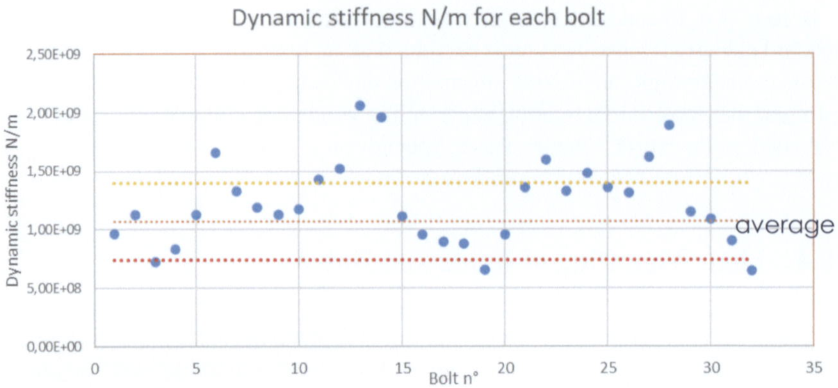

Fig. 10.11 Dynamic stiffness of the tested element

In this particular case, it is possible to link the change in dynamic stiffness to the number of cycles performed, and to monitor the behaviour of prestressed fasteners. The tension force on the fastener had been measured and was close to 320 kg. A phenomenon of fatigue is a measurable reality for this type of equipment (Fig. 10.12).

After maintenance operations, the figure below shows the evolution of the dynamic stiffnesses of some of the robot fixings on the concrete slab (Fig. 10.13).

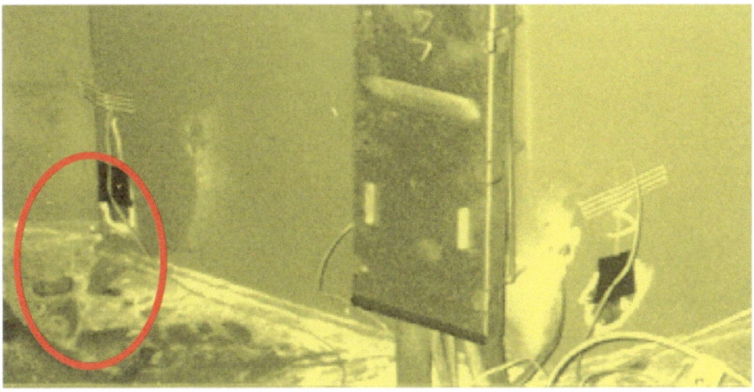

Fig. 10.12 Instrumented robot foot. *Source* Rincent BTP France

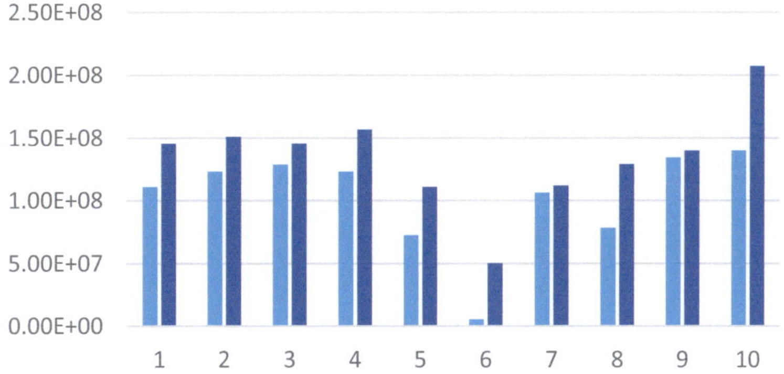

Fig. 10.13 Stiffnesses values after additional tension

Chapter 11
Tension Force Limit of an Element Fixed to a Support

Abstract Another use was to test building cladding panels without reaching the rupture. This chapter describes the procedure and how to use the results.

11.1 Tests Conditions

The stability of building facade plates with respect to the forces generated by wind circulation is a question that is regularly asked, whether these plates are fixed or glued. The immediate requirement following this request is not to break the facade plaques. The aim is therefore to find the tension force limit without causing the rupture (Figs. 11.1 and 11.2).

11.2 Procedures

The procedure consists of carrying out at least three simultaneous tests in which the static and dynamic stiffnesses are calculated. As indicated in Chap. 5, the ratio of Dynamic Stiffness to Static Stiffness can be calculated. Our patent n° EP1 763 667B1 in 2008: Method and Device for determining the tensile breaking strength of an element fixed to a support is no longer active and formalize this method (Appendix L). The following figure explains the procedure, the three first level led to calculate Rd/Rs to utilise this ratio to calculate Rs4 from the value Rd4. For the fourth point, this ratio permits to calculate the static stiffness from the dynamic stiffness and to draw the tangent at this point. The next step is to extend the curve and extrapolate the tension limit value (Figs. 11.3, 11.4 and 11.5).

The previous figure shows the tests performed on 42 façade plates on 6 locations to identify the quality of the fastening.

In the case of a large number of bonded plates, the procedure was as follows:

- Dynamic tests performed at the centre of the plates
- Calculation of dynamic stiffnesses

J. J. Rincent, *Anchors—Tests Procedures and Vibratory Analysis*,
https://doi.org/10.1007/978-981-96-3777-5_11

Fig. 11.1 Stone plaque fixed to a concrete wall. *Source* Rincent BTP France

Fig. 11.2 Example of a building with stone facing. *Source* Rincent BTP France

- Creation of homogeneous stiffnesses families
- Carrying out static tensile tests simultaneously with dynamic tests on samples of each family.

This type of test can be carried out on tiles in shopping centres and airport halls, and on all elements bonded to structural substrates (Fig. 11.6).

The following example was performed on the roof of a digester which contains methane under pressure. The concrete was poured on a concrete prefabricated structure. These tests were used to organize the manutention (Fig. 11.7).

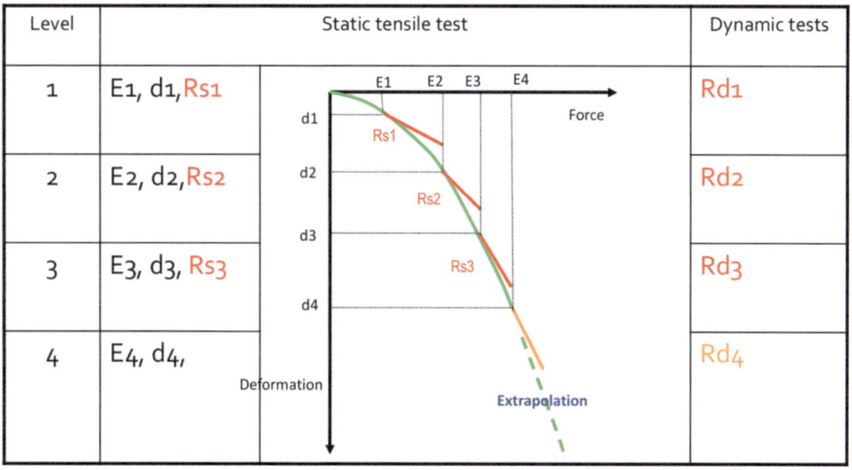

Fig. 11.3 Static and dynamic tests, extrapolation of limit tension force

Fig. 11.4 Device enabling simultaneous static and dynamic tests. *Source* Rincent BTP France

Fig. 11.5 Dynamic tests. *Source* Rincent BTP France

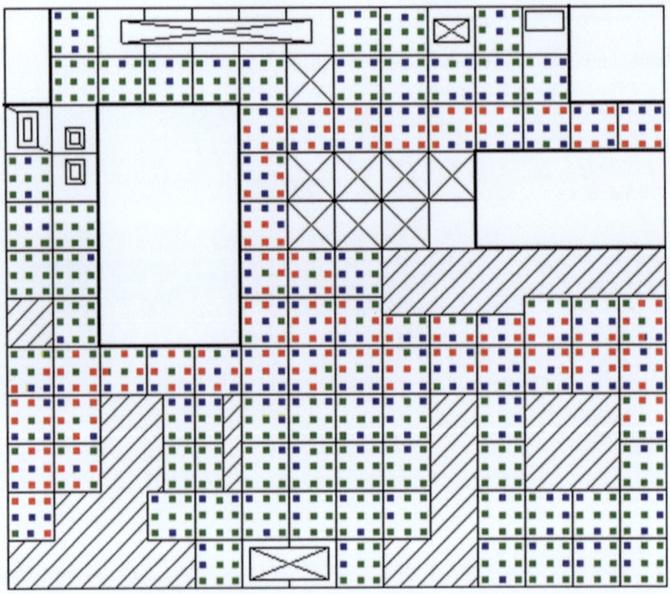

Fig. 11.6 Dynamic stiffnesses map

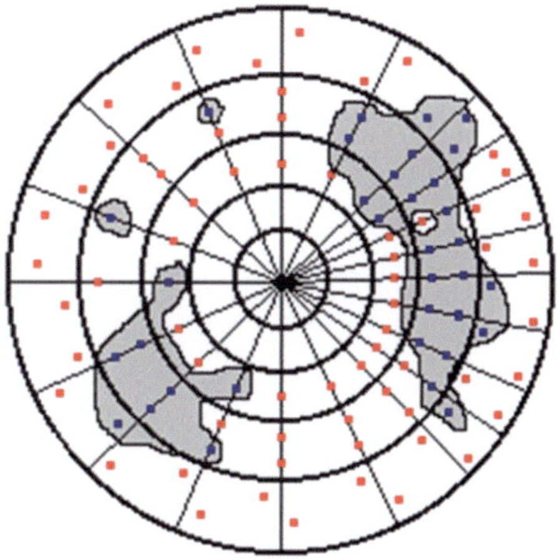

Fig. 11.7 Map of unstuck parts of the concrete dome

Chapter 12
Micro Piles

Abstract The method was used for micro piles to determine the mobilizable tension force. The example presented combines dynamic and static tests.

12.1 Current Tests

In the case of micro piles with a reinforcement consisting of a petroleum tube, it is possible to weld a plate to a sleeve to carry out the tests. It is necessary to fix the 3D geophone rigidly, not with grease as on the photo (Fig. 12.1).

In the same way, total length, free length and micro pile cross section can be calculated. The curve is identical to the previous curves, the Rincent BTP Technologies equipment have a presentation slightly different (Fig. 12.2).

The dynamic stiffnesses are calculated by the same way (Fig. 12.3).

On the chosen example, the stiffnesses of micro piles 6 and 12 correspond to the vibratory response of the device screwed into the upper part of the micro pile and are therefore not the stiffnesses of the micro pile itself.

The three exploitable stiffnesses, can be gathered with the three-creep load values T calculated from the static tests. (Appendix M).

The dynamic stiffnesses 6 and 12 are abnormally low because the geophone was not fixed in accordance with the procedure consisting of gluing a rigid interface into which the 3D geophone support is screwed. The three values used are those for micro pile numbers: 8, 13 and 31 (Fig. 12.4 and 12.5).

12.2 Soil Skin Friction Scaling

Tensile tests on micro piles are carried out in the same way as on a tension ground anchor.

The use of the method described in this book in the field of micro piles is quite rare. However, in exceptional cases where a complete detailed diagnosis of a batch

J. J. Rincent, *Anchors—Tests Procedures and Vibratory Analysis*,
https://doi.org/10.1007/978-981-96-3777-5_12

Fig. 12.1 Micro pile tests. *Source* Rincent BTP France

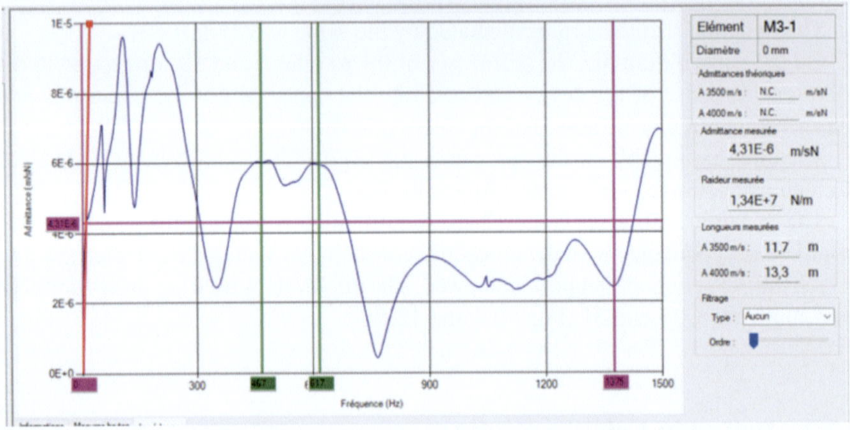

Fig. 12.2 Micro pile response

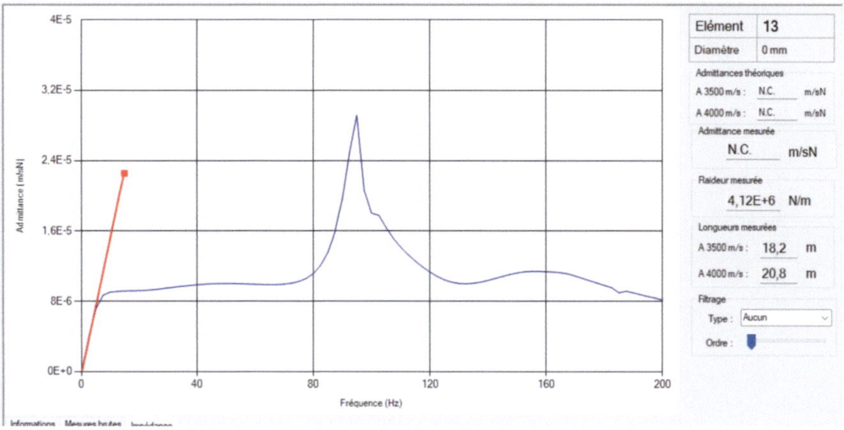

Fig. 12.3 Micro-pile stiffness

Fig. 12.4 Dynamic Stiffnesses square root table and creep load values

Micro pile n°	Dynamic stiffness N/m	(Dynamic stiffness)1/2	creep load tons
6	4,52E+05		
8	6,99E+06	2644	15,1
12	5,05E+05		
13	8,65E+06	2941	20,7
31	9,65E+06	3106	23,8

Fig. 12.5 (Dynamic stiffness)$^{1/2}$ function of force

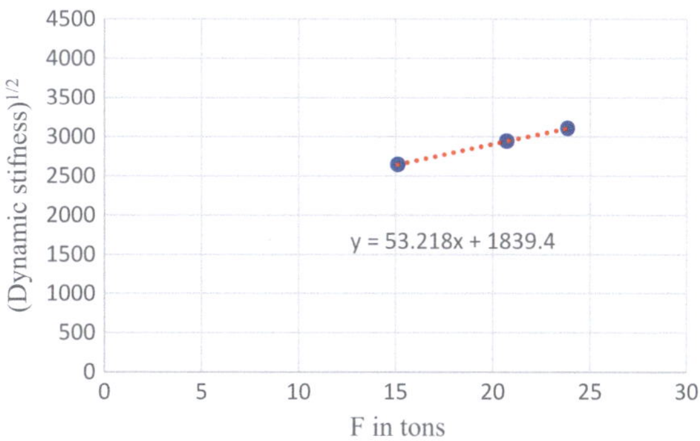

of micro piles is required, the method immediately becomes an unrivalled technical contribution.

In the same way as for tie rods, the dynamic tests are calibrated with static tests. The first graph concerns a micro pile tested in tension with dynamic vibration analysis tests (Fig. 12.6).

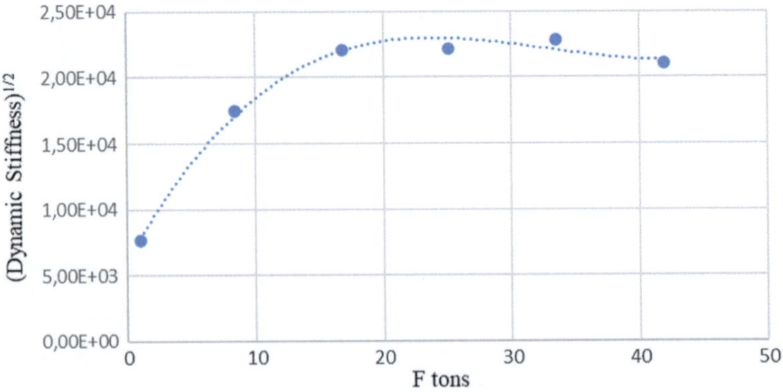

Fig. 12.6 Square root of the dynamic stiffness function of the tensile force

The curve obtained showing the relationship between the square root of the stiffness and the tension forces is this type (Fig. 12.7).

This result is quite similar to those obtained with large-diameter tie rods. For a micro pile where failure is obtained, the curve passes through a maximum. The first part of the curve before failure allows us to obtain a relationship between the square root of the dynamic stiffness and the tensile force. The formula changes for each type of micro-piles, the B value is slightly different due to the micro pile methodology which use less or more cement grout injected (Fig. 12.8).

Numerous calibration tests were carried out on a linearly extended site. At one point on the site, the tests produced the formula $y = 176x + 11{,}406$, i.e. close to the previous tests. This formula was used to define the critical creep tension of the micro pile:

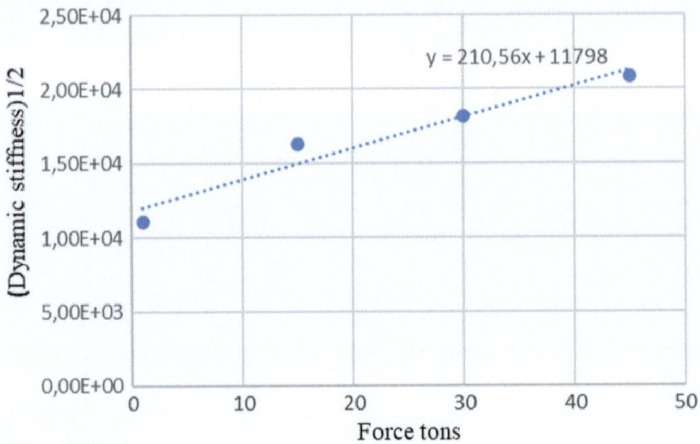

Fig. 12.7 Example of $y = Ax + B$ value

Fig. 12.8 Forces calculated for 3 types of micro-piles

Micro-piles	(Dynamic stiffness)1/2		calculated
Type 1	Type 2	Type 3	force tons
1,11E+04	1,48E+04	4,77E+03	1
		1,29E+04	8,4
1,63E+04			15
		1,88E+04	16,8
	1,79E+04		21,4
	1,84E+04		23,2
		1,93E+04	25,1
1,81E+04			30
		2,17E+04	33,5
		2,02E+04	41,9
	2,16E+04		42,7
2,08E+04			45

Micro pile 1 2.28E08 N/m 20.8 tons
Micro pile 2 2.97E08 N/m 33.3 tons

And then on these two micro piles, static tension tests have been performed, the results obtained for the critical creep tensions were 19.6 tons for the first and 32.2 tons for the second. On this site, 210 micro piles were tested to determine the force that could be applied to these foundations.

Chapter 13
Other Applications

Abstract The method has been used for soils nailing, embankment reinforcement, historic monument. An example of use of our database is given.

There are many possible applications, for example we have used the measurement of dynamic stiffnesses to monitor and validate the hardening of the grouting of the extrados of the voussoirs in the Channel Tunnel. Here we present a few examples that show the diversity of uses of the method.

13.1 Soils Nailing

Nailing is a method of reinforcing the stability of a slope that is carried out during the earthworks on the slope (Fig. 13.1).

On this structure, the question posed was to study the loading of passive anchors during earthworks. After calibration on one of the nails, the calculated tension forces varied from 0 to 9 tonnes (Fig. 13.2).

In the following example, the aim was to check the functioning of the nails on an old structure:

- Confirm the lengths (broken—no broken)
- And dynamic stiffnesses (Fig. 13.3).

The dynamic stiffnesses measured are:

n°	Stiffnesses N/m
48	9.59E+07
49	1.22E+08
50	1.97E+08
51	3.51E+08

J. J. Rincent, *Anchors—Tests Procedures and Vibratory Analysis*,
https://doi.org/10.1007/978-981-96-3777-5_13

Fig. 13.1 Soil nailing. *Source* Rincent BTP France

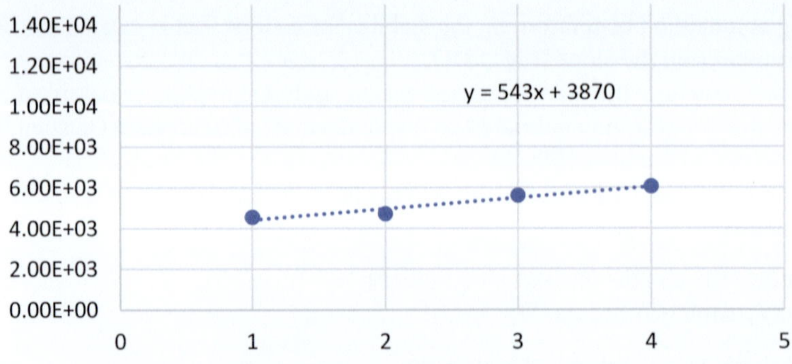

Fig. 13.2 Square root of the stiffness function of the force

And the square root of the stiffness as a function of the altimetric position of the nail gives the following graph which confirms that it operates in accordance with slope stability calculations (Figs. 13.4 and 13.5).

After dismantling the metal section of a lock gate, dynamic tests were carried out to check the nails which reinforced concrete mass.

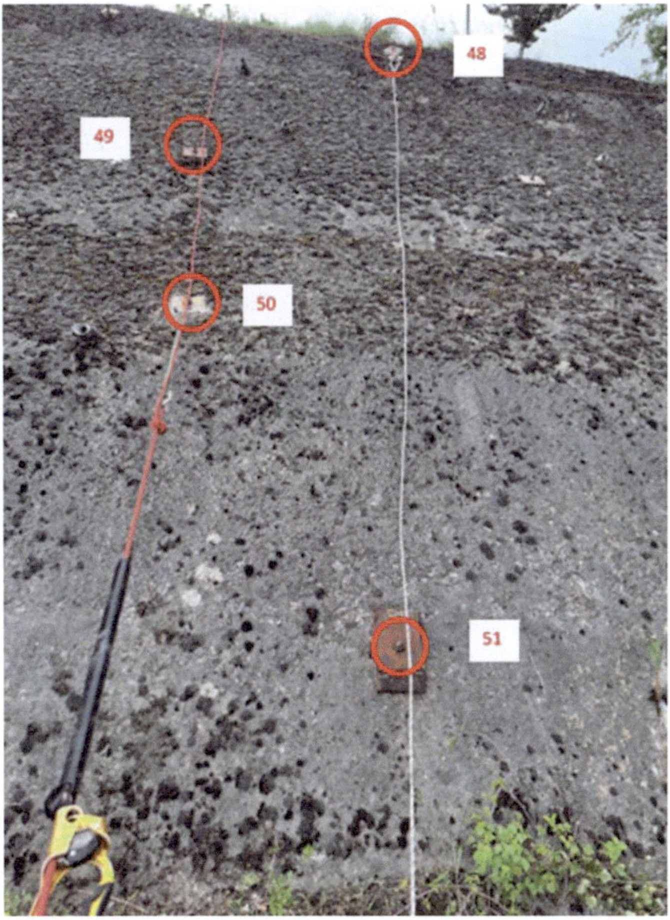

Fig. 13.3 Ancient soil nailing reinforcement. *Source* Rincent BTP France

The following figure shows tests carried out in the mountains on rocky cliff reinforcements. It should be remembered that in the case of a rock mass, the operation of reinforcements follows the rules of rock mechanics and not soil mechanics (Fig. 13.6).

13.2 Embankment Reinforcement

The tests were carried out on February 2012; their purpose was to provide a non-destructive diagnosis of the nails tested, in terms of length, dynamic stiffness and estimated tension force (Fig. 13.7).

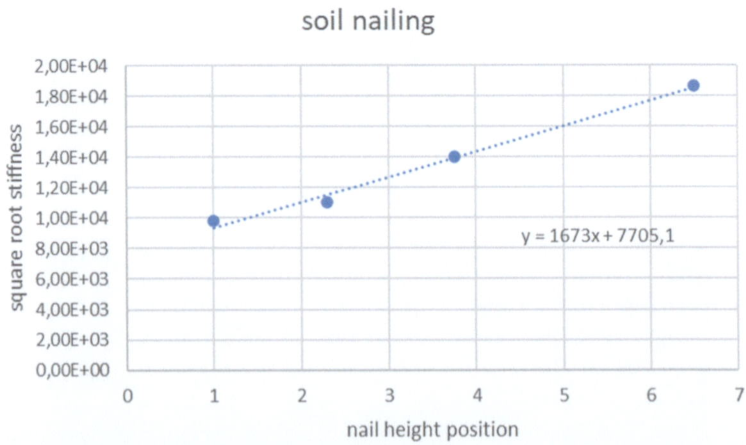

Fig. 13.4 Square root stiffness function of height position

Fig. 13.5 Port lock-tests on reinforcement nails. *Source* Rincent BTP France

Fig. 13.6 Cliff reinforcement. *Source* Rincent BTP France

Here too, the results must be interpreted taking into account the rocky nature of the site. will take two columns of results that seem anachronistic if these results are analysed as for a soil retaining wall.

level	Force in	
5	tonnes	20,4
4	18,1	14,5
3	9	3,1
2	2,6	0,5
1	0	0

Fig. 13.7 Reinforcement bars. *Source* Rincent BTP France

The thrust starts at level 2 and increases towards the top of the wall because at level 2 the space occupied by the soil fill is reduced. At level 4 the thrust increases sharply as this is the start of a road embankment. At the foot of the wall the dynamic stiffness is strong and corresponds to the contact of the wall with the rock in place and the fill thrust is null. Interpreting the results from a rock site is often complicated, especially in the absence of information.

13.3 Historic Monuments

Our test method is sometimes used on specific structures, in this case the Château de Chambord built 5 centuries ago and the reinforcement of a staircase with metal bars (Fig. 13.8).

13.4 Data Base

On one of the current sites, the client had asked for a diagnosis of the tie rods without initially relying on static calibration tests. The type of rebar, the length of the tie, its diameter and the size of the wall are the elements that guide the choice of a formula.

Three formulas close to this configuration are:

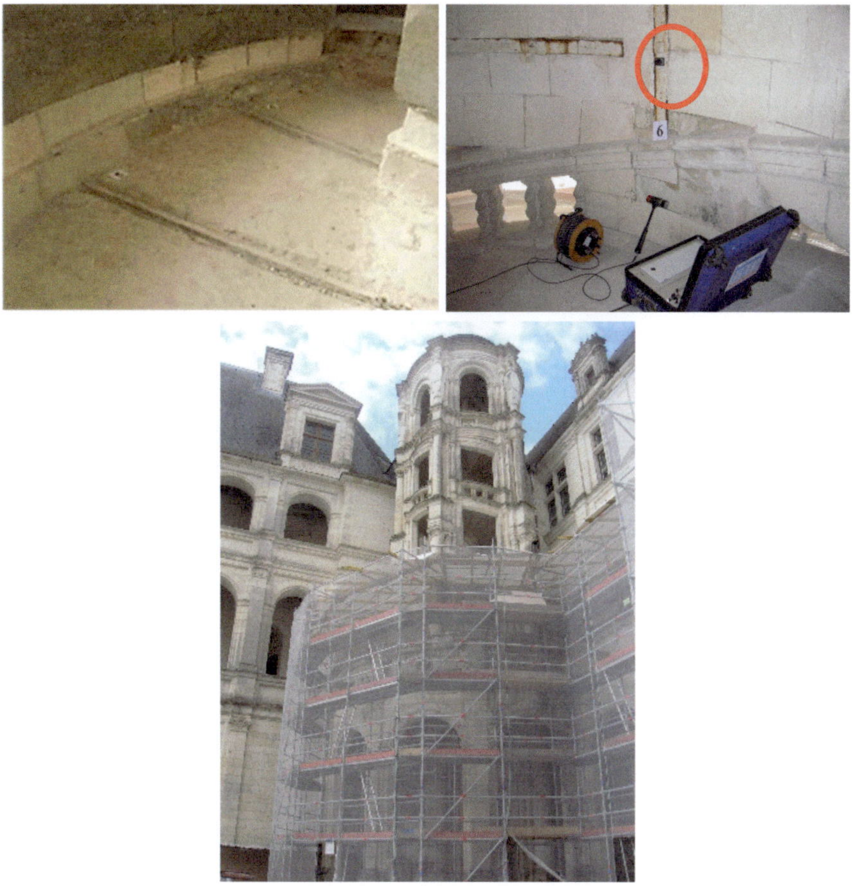

Fig. 13.8 Castle staircase. *Source* Rincent BTP France

$$\begin{aligned}&\text{Pk } 24{,}800 \ \text{y} = 2280\text{x} + 2328\\&\text{Pk } 35{,}300 \ \text{y} = 2805\text{x} + 2320\\&\text{Pk } 28{,}700 \ \text{y} = 2813\text{x} + 2247\end{aligned}$$

When the curves were analysed, a stiffness calculation revealed a value of 4.64E6, i.e. taking the square root of the stiffness 2154, which may correspond to the value of B with an order of size very close to the previous formulae.

The two A values de 2805 and 2810 were an element of the choice and in order not to be too optimistic about the tension values of the tie rods, the following formula has been chosen:

$$\text{Y} = 2810\text{x} + 2154$$

Several months later, two static tests were carried out on tie rods 37 and 77, for which the calculated dynamic stiffnesses were as follows:

The static tests on tie rod 77 show that the internal tension is slightly less than 10 tonnes (Figs. 13.9 and 13.10).

For the tie rod 37, the detachment of the support plate from the wall was recorded by the operator at a value of 10 tonnes and confirmed by the dynamic tests (Fig. 13.11 and 13.12).

In view of these results, the formula is as follows:

$$Y = 2330x + 2154$$

The tension forces of tie rods 37 and 77 are respectively: 10.15 tonnes and 9.71 tonnes.

The two conclusions are that it would have been preferable to choose the first formula and that, despite everything, the accuracy of the results is linked to the static calibration tests.

37	6,17E+08			77	5,93E+08		
	6,55E+08				5,95E+08		
	6,60E+08				6,12E+08		
	6,73E+08				6,59E+08		
	6,52E+08				6,55E+08		
	6,92E+08				6,48E+08		
	6,67E+08				6,20E+08		
	6,91E+08	6,66E+08	8,42t		6,34E+08	6,27E+08	8,14t

Fig. 13.9 Result table of dynamic stiffness N/m

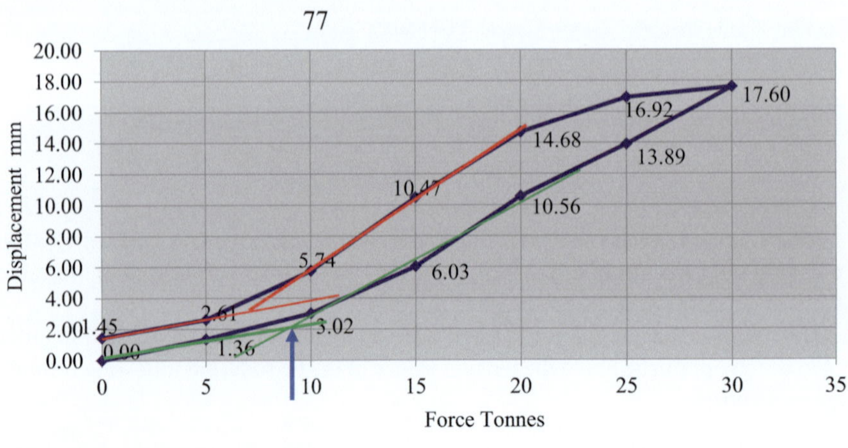

Fig. 13.10 Curve Force deformations

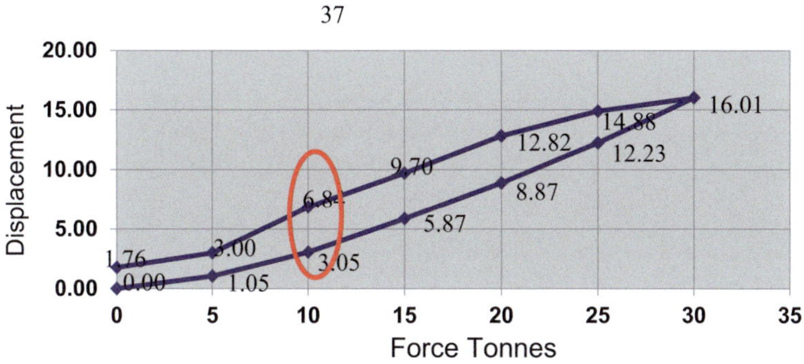

Fig. 13.11 Curve force deformation

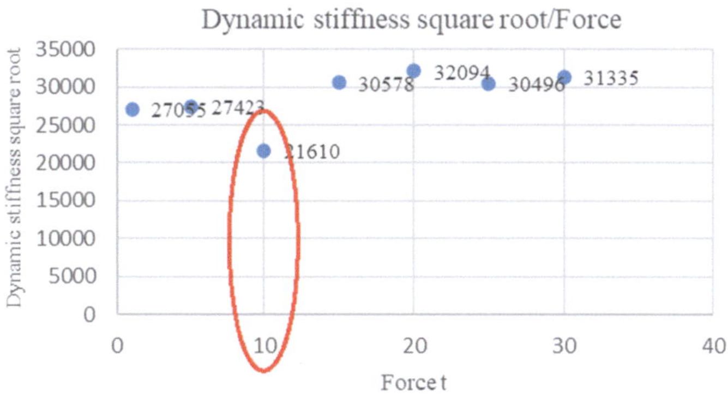

Fig. 13.12 Determination of the tension force (Appendix N)

Chapter 14
Test Procedure and Results

Abstract This chapter gathers the elements of a test procedure for the dynamic tests, for the simultaneous dynamic and static tests and interpretations.

This paragraph describes the impulse test on anchors pre tensioned or not. It defines the terms used and the parameters measured, specifies the characteristics of the equipment, lays down the test procedure and specifies the results to be presented.

We have taken the procedures we currently use for dynamic tests, static tests and simultaneous tests.

14.1 Dynamic Test

14.1.1 Definitions

For this test, the following definitions are applied:

- Mobility or mechanical admittance: Ratio between the particle velocity V of a vibrating point at the head of the anchor and the force F applied at the head of the anchor during the test.
- Mechanical impedance: Inverse value of the mechanical admittance.
- Dynamic stiffness of an anchor at low frequency: Ratio between the force F applied at the head of the pile and the displacement of the head of the pile at the start of the test.
- F is the applied force, expressed in newtons.
- V is the particle velocity of a vibrating point at the head of the anchor, expressed in metres per second.
- V_b is the celerity of plane waves in the anchor with the cement grouting, expressed in metres per second.
- ρ_b is the density of the concrete in the pile, expressed in kilograms per cubic metre.

J. J. Rincent, *Anchors—Tests Procedures and Vibratory Analysis*,
https://doi.org/10.1007/978-981-96-3777-5_14

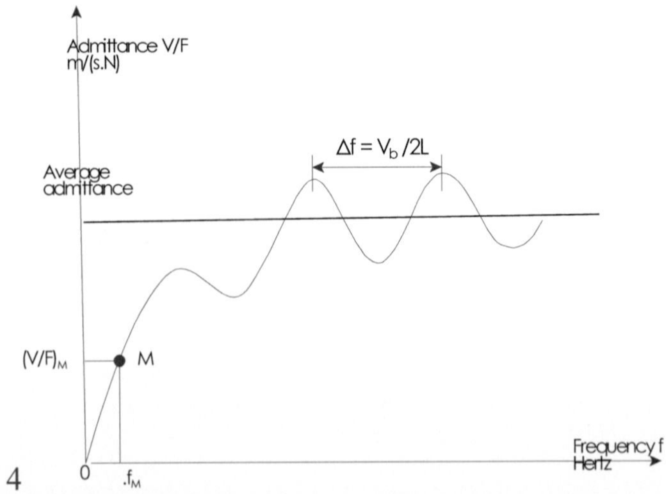

Fig. 14.1 V/F function of frequency

14.1.2 Results

The values obtained are as follows:

- The anchor dynamic stiffness R_d at low frequency is expressed in newtons per metre.
- V/F is the mobility or mechanical admittance expressed in metres per Newton-second.
- $(V/F)_M$ is the measured mechanical admittance or characteristic mobility, expressed in metres per second-newton.
- ΔF is the frequency difference between two successive maxima or minima of the signal amplitude in the frequency domain, expressed in Hertz (Fig. 14.1).

14.1.3 Elements of the Acquisition Chain

Equipment list:

- a power supply needed to operate the acquisition chain,
- a hammer fitted with a force transducer to produce and measure the intensity of the impact,
- a velocity transducer and its coupling product: geophone or accelerometer with its integrator,
- the fixing must be rigid; a support can be glued to the metal plate supporting the element to be tested and the geophone support is screwed into this support,
- equipment for acquiring, displaying, recording and processing fast signals,

- equipment for reproducing the recorded and processed signals.

14.1.4 Procedure

The following information is required to prepare and carry out the test:

- identification of the anchors being tested n°, location.
- the characteristics of the anchor.

The purpose of preparing the top of the anchor is to:

- facilitate access to the head.
- define two areas, one to stick the geophone 3D.
- and the other for the hammer strike (Fig. 14.2).

Fig. 14.2 Example of 3D geophone mounting. *Source* Rincent BTP Recife

Systematically take a photo of the head of the tie rod. This operation is a way of storing information specific to the tie rod tested, for example the presence of cracks in the wall or abnormal bearing plate positions, etc.…

Place the receiver, geophone or accelerometer, on the metallic plate of the anchor head, interposing a coupling system to ensure reception of the mechanical waves.

Check that external sources of vibration do not cause any significant disturbance. In this case, several keystrokes can be added together to improve signal quality by reducing the effect of parasitic vibrations.

Use the hammer to produce a mechanical shock at the head of the pile, directed parallel to the anchor axis, on the surface prepared.

Simultaneously measure the amplitude of the force applied at the moment of impact and the particle velocity.

Visualise the signal.

Repeat the acquisitions to reach the number 8 and record them.

14.1.5 Results

Lengths

The vibratory response of the element under test is used to calculate the distance between two peaks or two troughs in order to calculate the length of the element that generated this response, for example the length of the free part and the total length.

$$L = V_b/2\Delta F$$

For V_b value, velocity of the waves in the anchor, in the absence of data, take $V_b =$ 4000 m/s.

On the graph (Fig. 12.1), determine the variation in frequency ΔF between two successive maxima or minima in the admittance of the wave signal.

Dynamic stiffness

The stiffness of the soil-anchor system from the inverse of the slope at the origin of the V/F curve in the frequency domain:

$$R_d = 2\pi . F_M/(V/F)_M$$

Equivalent cross-section

The cross-section of a tie rod is rarely circular, particularly in materials with low mechanical properties. Which is why this section is called equivalent, the surface can be correct without being circular.

The characteristic admittance (V/F) corresponding to the mean value of the successive minima and maxima of the admittance.

The theoretical value of the characteristic admittance is:

$$(v/F) = (p_b.V_b.A) - 1$$

In the absence of data on the density of the ground anchor with its cement grout take:

$$P_b = 2400\ kg/m^3$$

Calculate the average admittance V/F in the frequency domain.

14.1.6 Test Report

The test report shall include the following minimum information:

- the name of the firm that carried out the test and the name and visa of the person responsible for the test,
- the file number and date of the test,
- the location and marking of the anchor tested,
- the geometric characteristics of the anchor,
- the graphic representation of the admittance V/F in the frequency domain,
- the calculated lengths L of the different parts of the tested element, the uncertainty about the result is linked to the uncertainty about the velocity in the element,
- the dynamic stiffness of the element,
- and the equivalent cross section.

Observations relating to the performance of the test as well as incidents and operating details not provided for in the standard and likely to have an influence on the results.

14.1.7 Verification of the Equipment

Measuring instruments and equipment are managed in accordance with the requirements laid down by the national metrology office.

The reception transducer and the force measurement transducer are calibrated with their measurement chain every six months at the most.

The calibration's expiry date and the name of the authorised body that carried it out appear on the report.

Before any work is carried out, the correct operation of the equipment is checked in the laboratory.

14.2 Static Tests

The static test is standardised, but it should be remembered that tests carried out at height involve a limited runtime.

The aim is not to require complicated tests on 40-year-old tie rods with a risk of failure that no one will ultimately realise. The aim is to find out the internal tension of the tie rod and, if possible, elsewhere than at the foot of the wall.

14.2.1 *Particular Cases*

When the value of the tension forces of two elements are known precisely, it is then possible to calculate the values a and b of the formula relating the square root of the dynamic stiffness and the tension force.

However, in most cases for older ground anchors, the tension forces are not known.

14.2.2 *Simultaneous Static and Dynamic Tests*

The first step is to design the equipment so that the two tests can be carried out simultaneously.

On old elements, it is advisable to have levels of limited amplitude, e.g. 5 tonnes, which means that n levels should be made and continued beyond the point where the wall support plate becomes detached, which is an indication that the internal tension force has been exceeded.

For each level 8 dynamic tests acquisitions are done, for loading and unloading.

14.2.3 *Results*

The static test is classical and determine the detachment force and the curve force—deformation (Fig. 14.3).

Example of a test sheet:

The calculation of the dynamic stiffnesses leads to the following curve, which highlights the internal tension force of the ground anchors as explained above (Figs. 14.4, 14.5 and 14.6).

The internal tensile force at the tie rod can be determined in two ways:

- From the direct tensile test on unloading, so as not to include the effort required to lift the plate from the wall that has been in place for several decades.
- From the dynamic stiffnesses, which are reduced when the tensile force exceeds the internal tension of the tie-rod.

Fig. 14.3 Simultaneous static and dynamic tests—example of a device. *Source* Rincent BTP Recife

14.3 Final Objectives

The objective is to define the formula which, based on the dynamic stiffnesses, calculates the internal tension forces of the tie rods.

Another objective is to identify the ground anchors that can be re-tensioned and those that will not withstand this operation. The stability of the dynamic stiffnesses combined with information from the static tests are the means of identifying these tie rods.

Static Test							
	load (tf)		duration (min)	deformation			
				DEFLEC. 01	DEFLEC. 02	average	Dynamic tests
	1,0		1 min				1 a 8
loading	5,0		1 min				
			5 min				
			10 min				9 a 16
	10,0		1 min				
			5 min				
			10 min				17 a 24
	15,0		1 min				
			5 min				
			10 min				25 a 32
	20,0		1 min				
			5 min				
			10 min				33 a 40
	25,0		1 min				
			5 min				
			10 min				41 a 48
	30,0		1 min				
			5 min				
			10 min				49 a 56
	35,0		1 min				
			5 min				
			10 min				57 a 64
	40,0		1 min				
			5 min				
			10 min				65 a 72
unloading	35,0		5 min				73 a 80
	30,0		5 min				81 a 88
	25,0		5 min				89 a 96
	20,0		5 min				97 a 104
	15,0		5 min				105 a 112
	10,0		5 min				113 a 120
	5,0		5 min				121 a 128
	0,0		5 min				129 a 136
Observations	detachment force						

Fig. 14.4 Test sheet. *Source* Rincent BTP Recife

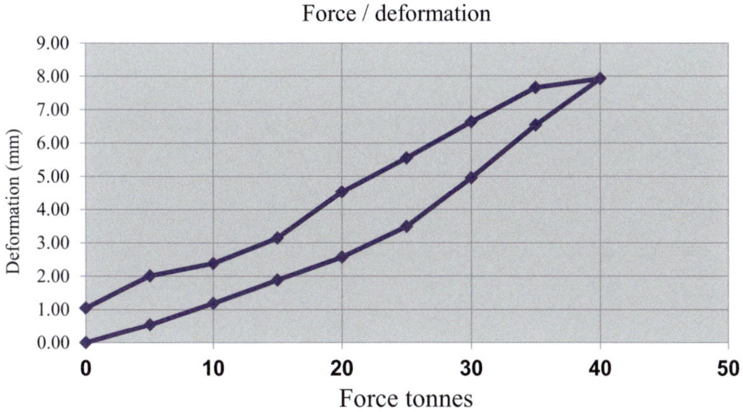

Fig. 14.5 Curve force–deformation example

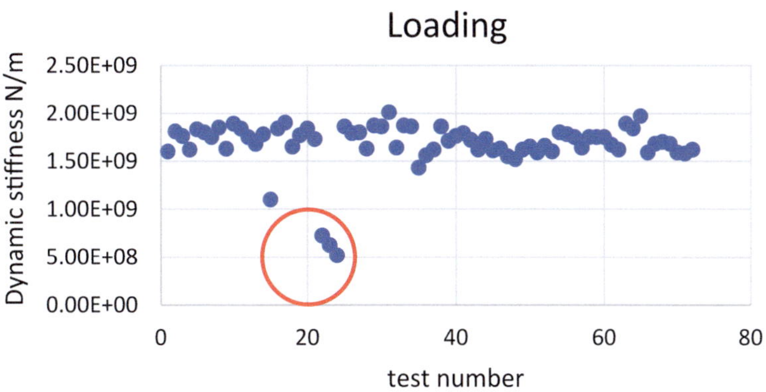

Fig. 14.6 Dynamic stiffness results

Chapter 15
Conclusions

This year 2024 we have tested 750 ground anchors; it is more of 6000 dynamic tests and more of 50 static tests. We have met numerous types of tie rods existing for 3 or 4 decades.

Each new project involving old ground anchors is a highly formative adventure that leads to the re-discovery of methods that have sometimes become obsolete. The aim of these tests is to provide managers of retaining structures with the information they need to diagnose and support maintenance.

Increasing the durability of these structures is at the heart of the approach. Putting the tie rods back under tension is one of the objectives, which needs to be clarified on the basis of the tests carried out.

The practice of this test over two decades has enabled us to establish an execution framework with a view to providing:

- the internal tensions of the tie rods,
- total lengths,
- free lengths,
- and cross-sections of the tie rods with their grouting.

These results provide a diagnostic tool for carrying out maintenance operations.

There are many areas of use:

- active ground anchors, with bars or strands,
- passive ground anchors,
- reinforcement using nails,
- prestressed dowels, etc.

The aim is to provide information that will lead to the definition of operations to be carried out to increase the durability of the structure.

This test is already being used to determine tension losses as a function of time, making it possible to monitor the development of the phenomenon and carry out re-tensioning as required.

J. J. Rincent, *Anchors—Tests Procedures and Vibratory Analysis*,
https://doi.org/10.1007/978-981-96-3777-5_15

Re-tensioning is a subject that needs to be formalised further by defining procedures and carrying out technical and economic analyses.

Appendix A

Scheme extract from the European patent EP 1 761 750 B1.

J. J. Rincent, *Anchors—Tests Procedures and Vibratory Analysis*,
https://doi.org/10.1007/978-981-96-3777-5

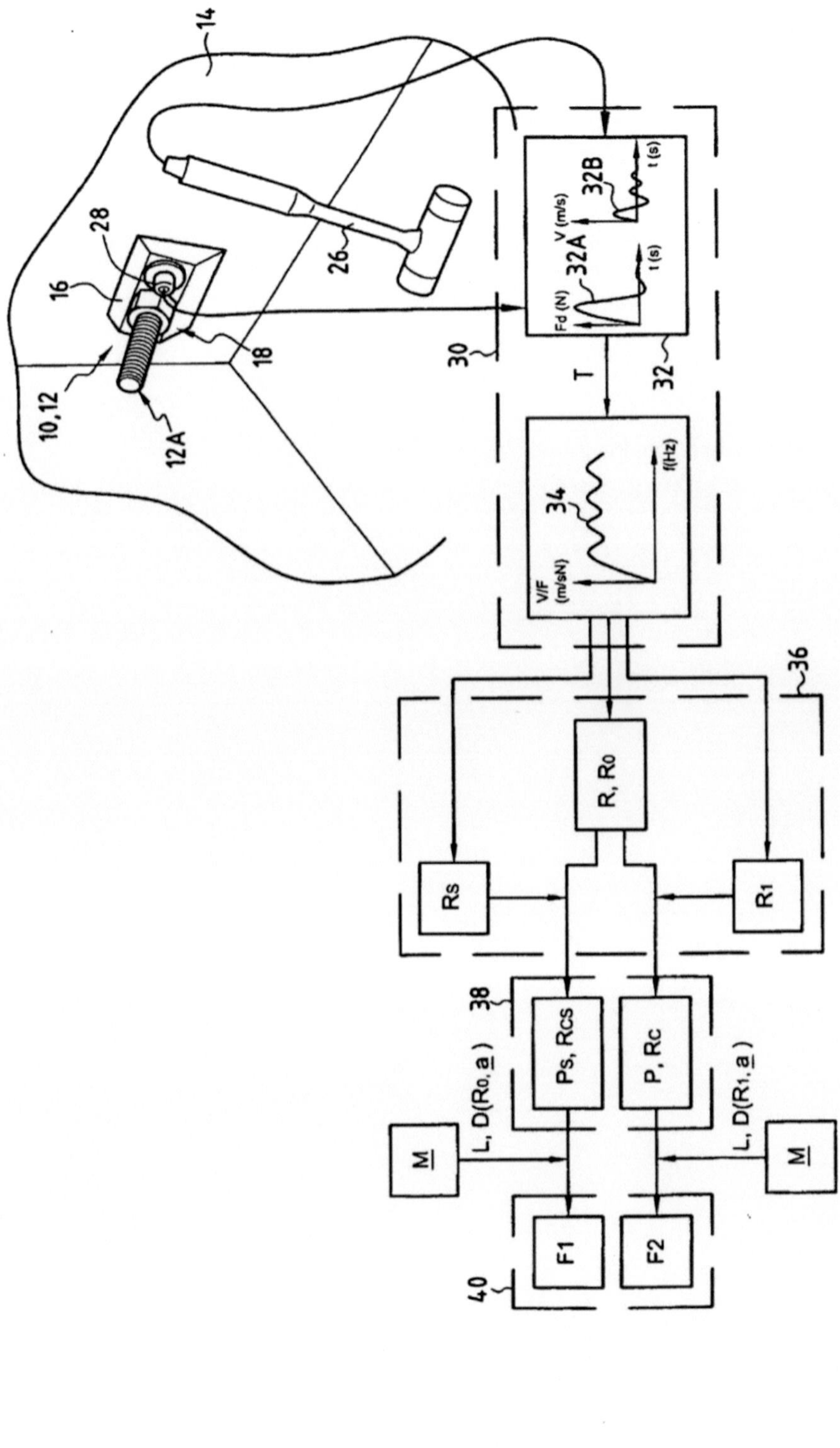
14
16
28
10,12
12A
18
26
30
32
Fd (N)
32A
t (s)
V (m/s)
32B
t (s)
T
V/F
(m/sN)
34
f(Hz)
36
R, Ro
Rs
R1
38
Ps, Rcs
P, Rc
L, D(Ro, a)
L, D(R1, a)
M
M
40
F1
F2

Appendix B

Stiffnesses Central part of the retaining wall.

J. J. Rincent, *Anchors—Tests Procedures and Vibratory Analysis*,
https://doi.org/10.1007/978-981-96-3777-5

411	5,73E+08	412	4,87E+08
	5,57E+08		-
	4,59E+08		4,72E+08
	5,76E+08		4,86E+08
	4,90E+08		3,44E+08
	5,64E+08		5,12E+08
	5,57E+08		5,19E+08
	5,19E+08		6,80E+08
424	1,24E+09	425	8,25E+08
B	1,15E+09		2,54E+08
C	1,04E+09		7,05E+08
D	1,17E+09		5,97E+08
E	1,14E+09		6,22E+08
F	1,06E+09		6,56E+08
G	1,15E+09		5,40E+08
H	1,19E+09		2,58E+08
437	2,26E+09	438	4,30E+07
	1,37E+09		3,88E+08
	1,88E+09		1,44E+08
	1,35E+09		4,72E+08
	1,52E+09		8,83E+08
	1,44E+09		1,15E+09
	1,60E+09		1,02E+09
	1,70E+09		5,73E+08

450	1,53E+08	451	1,59E+08
	1,85E+08		1,70E+08
	3,04E+08		1,03E+08
	1,35E+08		1,29E+08
	1,67E+08		8,40E+07
	2,15E+08		8,39E+07
	1,73E+08		9,28E+07
	2,22E+08		1,06E+08
463	1,52E+09	464	1,54E+08
	1,51E+09		2,10E+08
	1,53E+09		1,13E+08
	1,54E+09		2,21E+08
	1,39E+09		7,50E+07
	1,59E+09		2,45E+08
	1,39E+09		2,63E+08
	1,45E+09		2,82E+08
476	1,32E+09	477	1,95E+09
	1,44E+09		2,21E+09
	1,65E+09		1,85E+09
	1,69E+09		2,40E+09
	1,54E+09		1,95E+09
	1,53E+09		2,31E+09
	1,42E+09		1,94E+09
	1,73E+09		2,19E+09

489	2,73E+09	490	3,40E+08
	2,91E+09		3,04E+08
	2,76E+09		3,17E+08
	3,23E+09		2,33E+08
	2,60E+09		4,49E+08
	2,59E+09		4,90E+08
	3,11E+09		-
	3,02E+09		1,70E+08
4102	7,08E+08	4103	9,84E+08
	1,17E+09		1,27E+09
	1,35E+09		1,60E+09
	1,29E+09		2,33E+09
	2,09E+09		1,15E+09
	1,02E+09		1,15E+09
	6,11E+08		2,67E+09
	-		2,58E+09
4115	9,33E+08	4116	2,31E+09
	1,20E+09		1,16E+09
	8,04E+08		1,07E+09
	1,08E+09		1,20E+09
	1,37E+09		2,68E+09
	1,87E+09		2,25E+09
	1,99E+09		2,29E+09
	2,01E+09		3,34E+09

4128	1,42E+09
	1,27E+09
	2,74E+09
	2,83E+09
	1,62E+09
	1,59E+09
	1,84E+09
	1,98E+09

4129	4,47E+08
	1,19E+08
	1,88E+08
	1,95E+08
	5,06E+08
	3,04E+08
	4,01E+08
	6,61E+07

4141	3,47E+09
	3,06E+09
	3,20E+09
	3,22E+09
	2,73E+09
	3,24E+09
	3,23E+09
	2,92E+09
4149	2,15E+09
	2,73E+09
	2,75E+09
	2,76E+09
	2,69E+09
	2,73E+09
	2,73E+09
	2,71E+09

4142	1,82E+09
	1,96E+09
	2,62E+09
	2,65E+09
	3,08E+09
	2,68E+09
	2,93E+09
	2,87E+09
4150	1,95E+09
	2,22E+09
	2,33E+09
	2,32E+09
	2,35E+09
	2,39E+09
	1,93E+09

Appendix C

Dynamic stiffnesses—Rainwater drainage system effect.

J. J. Rincent, *Anchors—Tests Procedures and Vibratory Analysis*,
https://doi.org/10.1007/978-981-96-3777-5

226	9,07E+08			238	6,66E+08		
	9,78E+08				7,60E+08		
	9,39E+08				7,13E+08		
	9,48E+08				7,52E+08		
	9,91E+08				6,85E+08		
	8,87E+08				7,20E+08		
	9,71E+08				7,33E+08		
	9,77E+08	9,53E+08	16,4t		7,27E+08	7,22E+08	13,5t
227	1,17E+09			239	9,13E+08		
	1,15E+09				8,72E+08		
	1,17E+09				8,81E+08		
	1,19E+09				9,24E+08		
	1,19E+09				7,99E+08		
	1,10E+09				8,85E+08		
	1,18E+09				8,88E+08		
	1,20E+09	1,18E+09	19,0t		9,27E+08	8,94E+08	15,7t
228	9,87E+08			240	7,35E+08		
bas	1,09E+09			bas	7,15E+08		
	1,10E+09				1,55E+09		
	1,03E+09				1,70E+09		
	1,03E+09				1,38E+09		
	1,17E+09				1,34E+09		
	9,65E+08				1,24E+09		
	1,05E+09	1,05E+09	17,5t		1,28E+09	1,25E+09	12,8t

Appendix D

Simultaneous tests Load and unload results T 253.

J. J. Rincent, *Anchors—Tests Procedures and Vibratory Analysis*,
https://doi.org/10.1007/978-981-96-3777-5

1	9,50E+08			20	1,06E+09	
	7,71E+08				9,79E+08	
	1,00E+09				9,08E+08	
	8,22E+08				8,88E+08	
	7,00E+08				7,87E+08	
	6,16E+08				9,47E+08	
	8,34E+08				9,34E+08	
	8,91E+08	8,36E+08			9,42E+08	9,33E+08
5	9,80E+08			25	9,20E+08	
	8,02E+08				9,90E+08	
	9,88E+08				9,40E+08	
	1,04E+09				9,70E+08	
	1,06E+09				1,00E+09	
	9,93E+08				8,08E+08	
	8,54E+08				8,50E+08	
	9,41E+08	9,66E+08			9,72E+08	9,40E+08
10	9,81E+08			30	8,84E+08	
	8,44E+08				1,00E+09	
	1,06E+09				8,56E+08	
	9,49E+08				9,85E+08	
	8,57E+08				9,85E+08	
	8,28E+08				1,02E+09	
	7,94E+08				8,80E+08	
	8,71E+08	8,88E+08			1,05E+09	9,59E+08
15	8,58E+08			35	1,00E+09	
	6,49E+08				8,55E+08	
	1,08E+09				9,72E+08	
	1,09E+09				1,00E+09	
	5,49E+08				1,06E+09	
	9,24E+08				9,03E+08	
	5,52E+08				9,64E+08	
	5,23E+08	7,69E+08			1,01E+09	9,75E+08

40	7,69E+08			15	8,70E+08	
	7,69E+08				9,26E+08	
	9,10E+08				6,69E+08	
	9,91E+08				1,01E+09	
	8,41E+08				7,80E+08	
	8,39E+08				8,65E+08	
	9,35E+08				9,21E+08	
	8,20E+08	8,52E+08			6,76E+08	8,39E+08
30	3,83E+08			10	6,73E+08	
	1,64E+08				8,07E+08	
	2,12E+08				7,86E+08	
	8,50E+08				7,62E+08	
	7,21E+08				9,44E+08	
	5,78E+08					
	9,41E+08				8,12E+08	
	9,46E+08	6,14E+08			8,73E+08	8,31E+08
25	9,26E+08			5	7,04E+08	
	7,61E+08				8,08E+08	
	9,15E+08				8,50E+08	
	5,60E+08				8,66E+08	
	4,34E+08				7,55E+08	
	5,85E+08				7,30E+08	
	4,10E+08				8,17E+08	
	1,57E+08	6,11E+08			7,22E+08	7,80E+08
20	6,39E+08			0	8,07E+08	
	9,07E+08				8,06E+08	
	9,95E+08				7,96E+08	
	8,90E+08				7,81E+08	
	8,43E+08				8,09E+08	
	8,66E+08				7,65E+08	
	1,03E+09				8,59E+08	
	9,62E+08	9,11E+08			7,53E+08	7,94E+08

Appendix E

Simultaneous tests T352.

J. J. Rincent, *Anchors—Tests Procedures and Vibratory Analysis*,
https://doi.org/10.1007/978-981-96-3777-5

1	8,91E+08	12	9,42E+08
	9,16E+08		1,04E+09
	9,68E+08		8,92E+08
	1,12E+09		6,82E+08
	1,01E+09		7,55E+08
	9,58E+08		7,39E+08
	9,71E+08		6,93E+08
	9,20E+08		8,98E+08
3	1,04E+09	15	1,85E+08
	9,25E+08		1,08E+08
	9,99E+08		2,51E+08
	9,13E+08		9,90E+07
	7,64E+08		5,73E+07
	9,76E+08		6,12E+08
	8,70E+08		1,01E+09
	1,02E+09		9,46E+08
6	7,59E+08	18	1,14E+09
	6,25E+08		9,31E+08
	8,63E+08		9,76E+08
	7,47E+08		9,43E+08
	1,12E+09		1,02E+09
	8,89E+08		1,00E+09
	5,91E+08		9,68E+08
	8,67E+08		9,45E+08
9	8,57E+08	21	1,04E+09
	9,47E+08		9,68E+08
	7,81E+08		9,82E+08
	8,94E+08		1,03E+08
	9,07E+08		9,95E+08
	1,32E+09		8,00E+08
	8,17E+08		8,00E+08
	9,46E+08		9,46E+08

24	9,38E+08	12	7,87E+08
	7,74E+08		9,99E+08
	8,11E+08		9,65E+08
	8,66E+08		1,24E+09
	8,72E+08		8,72E+08
	9,32E+08		9,42E+08
	8,15E+08		9,66E+08
	9,65E+08		7,20E+09
21	8,20E+08	9	9,48E+08
	7,38E+08		8,25E+08
	8,31E+10		8,57E+08
	8,38E+08		9,20E+08
	8,52E+08		8,93E+08
	8,32E+08		8,06E+08
	8,75E+08		8,30E+08
	9,20E+08		8,51E+08
18	8,98E+08	6	7,19E+08
	8,04E+08		9,15E+08
	9,75E+08		7,75E+08
	8,58E+08		9,29E+08
	8,70E+08		9,56E+08
	1,04E+09		8,54E+08
	9,19E+08		9,44E+08
	9,46E+08		8,21E+08
15	8,87E+08	3	7,74E+08
	9,53E+08		9,21E+08
	9,63E+08		9,79E+08
	1,02E+09		9,03E+08
	7,08E+08		9,23E+08
	1,17E+09		9,64E+08
	9,92E+09		8,37E+00
	8,86E+08		8,76E+08

		0	9,01E+08
			9,52E+08
			7,73E+08
			9,55E+08
			1,01E+09
			1,00E+07
			8,89E+08
			1,06E+09

Appendix F

Simultaneous tests T15

J. J. Rincent, *Anchors—Tests Procedures and Vibratory Analysis*, https://doi.org/10.1007/978-981-96-3777-5

1	7,85E+08		32	8,54E+08	
	6,97E+08			8,34E+08	
	6,53E+08			7,98E+08	
	6,50E+08			8,89E+08	
	7,48E+08			7,57E+08	
	7,06E+08			8,69E+08	
	6,67E+08			7,68E+08	
	7,03E+08	6,96E+08		8,80E+08	8,34E+08
8	7,51E+08		40	8,09E+08	
	8,16E+08			8,91E+08	
	7,67E+08			8,50E+08	
	7,93E+08			8,80E+08	
	7,15E+08			6,88E+08	
	7,60E+08			6,95E+08	
	8,63E+08			6,28E+08	
	8,42E+08	7,88E+08		8,48E+08	7,87E+08
16	7,31E+08		48	8,88E+08	
	7,38E+08			8,88E+08	
	8,46E+08			9,99E+08	
	8,37E+08			8,83E+08	
	7,87E+08			8,89E+08	
	7,28E+08			8,64E+08	
	7,18E+08			9,85E+08	
	7,63E+08	7,64E+08		8,72E+08	9,01E+08
24	8,85E+08				
	7,97E+08				
	8,98E+08				
	9,49E+08				
	8,11E+08				
	7,86E+08				
	8,04E+08				
	7,94E+08	8,40E+08			

Appendix G

Concrete dam T42.

J. J. Rincent, *Anchors—Tests Procedures and Vibratory Analysis*,
https://doi.org/10.1007/978-981-96-3777-5

	7,36E+09				
1,2t	1,34E+10		50t	7,45E+09	
	1,27E+10			8,67E+09	
	1,27E+10			8,96E+09	
	1,30E+10			9,39E+09	
	1,09E+10			9,71E+09	
	1,17E+10			9,95E+09	
	1,02E+10	1,19E+10		8,89E+09	
				8,86E+09	9,08E+09
10t	7,69E+09				
	8,17E+09		40t	8,89E+09	
	7,92E+09			7,38E+09	
	8,14E+09	7,98E+09		7,81E+09	
	8,59E+09			8,08E+09	
	8,29E+09			8,64E+09	
	8,70E+09			9,83E+09	
	7,46E+09	8,26E+09		9,38E+09	
		8,13E+09		9,45E+09	8,71E+09
20t	7,58E+09				
	7,24E+09		30t	8,83E+09	
	1,01E+10			9,54E+09	
	9,66E+09			8,52E+09	
	8,23E+09			9,38E+09	
	9,51E+09			9,26E+09	
	9,38E+09			9,53E+09	
	8,38E+09	8,79E+09		9,68E+09	
				1,03E+10	9,37E+09
30t	1,01E+10				
	9,52E+09		20	9,87E+09	
	9,54E+09			9,98E+09	
	1,02E+10			9,60E+09	
	9,54E+09			8,70E+09	
	1,08E+10			8,27E+09	
	9,85E+09			7,90E+09	
	1,03E+10	1,00E+10		8,04E+09	
				9,42E+09	8,98E+09
40t	9,11E+09		10t	8,46E+09	
	9,31E+09			8,83E+09	
	1,03E+10			9,42E+09	
	8,78E+09			8,81E+09	
	1,14E+10			1,06E+10	
	1,03E+10			9,72E+09	
	1,08E+10			8,45E+09	
	1,14E+10	1,02E+10		9,03E+09	9,05E+09
50	8,98E+09				
	9,62E+09		1,2t	7,93E+09	
	9,82E+09			7,36E+09	
	9,66E+09			8,51E+09	
	1,10E+10			8,02E+09	
	1,08E+10			8,60E+09	
	1,09E+10			8,06E+09	
	1,05E+10	1,02E+10		8,65E+09	
				8,89E+09	8,30E+09
55	9,65E+09				
	9,52E+09				
	9,72E+09				
	9,54E+09				
	9,42E+09				
	8,37E+09				
	9,91E+09				
	8,07E+09	9,37E+09			

Appendix H

Central part retaining wall.

J. J. Rincent, *Anchors—Tests Procedures and Vibratory Analysis*,
https://doi.org/10.1007/978-981-96-3777-5

504	4,76E+08	505	5,36E+08	507	2,10E+08	508	5,08E+11
	5,49E+08		4,85E+08		9,57E+07		6,57E+07
	4,47E+08		5,38E+08		3,06E+08		1,31E+08
	4,57E+08		5,99E+08		2,00E+08		9,57E+07
	5,73E+08		4,66E+08		3,19E+08		1,39E+08
	4,35E+08		4,67E+08		1,99E+08		1,16E+08
	3,89E+08		4,55E+08		4,88E+08		1,86E+08
	4,47E+08		6,17E+08		1,28E+08		5,45E+07
517	1,66E+09	518	8,95E+08	520	8,24E+08	521	8,74E+08
	1,14E+09		8,74E+08		9,81E+08		8,46E+08
	6,38E+08		8,99E+08		7,41E+08		8,06E+08
	2,33E+08		8,58E+08		8,06E+08		9,33E+08
	9,84E+08		1,04E+09		7,53E+08		6,78E+08
	1,14E+09		1,03E+09		7,28E+08		9,47E+08
	1,03E+09		9,78E+08		7,04E+08		9,19E+08
	1,00E+09		1,08E+09		8,38E+08		9,56E+08
530	2,84E+08	531	8,08E+08	533	1,31E+08	534	8,28E+08
	2,96E+08		7,34E+08		2,00E+08		6,34E+08
	2,50E+08		7,73E+08		1,84E+08		8,63E+08
	2,53E+08		8,01E+08		1,91E+08		9,38E+08
	1,07E+08		8,40E+08		2,56E+08		8,54E+08
	5,72E+08		8,25E+08		2,37E+08		8,75E+08
	4,48E+08		8,16E+08		6,64E+07		7,18E+08
	3,93E+08		7,10E+08		2,79E+08		8,09E+08
543	2,69E+08	544	1,14E+08	546	1,08E+08	547	7,54E+08
	8,32E+07		8,41E+07		1,48E+08		1,03E+09
	5,39E+08		9,04E+07		2,19E+08		1,14E+09
	5,25E+08		6,76E+07		1,89E+08		8,77E+08
	2,08E+08		5,48E+08		3,43E+08		9,69E+08
	5,49E+08		1,29E+08		1,22E+08		7,10E+08
	6,02E+08		1,26E+08		1,16E+08		1,04E+09
	3,23E+08		9,72E+07		2,45E+08		7,63E+08
556	1,25E+09	557	1,10E+09	559	1,90E+08	560	1,12E+09
	1,33E+09		1,06E+09		7,77E+08		1,13E+09
	1,04E+09		1,05E+09		9,98E+08		1,38E+09
	9,87E+08		1,11E+09		1,02E+09		1,26E+09
	8,39E+08		1,14E+09		8,47E+08		1,19E+09
	1,09E+09		8,77E+08		9,28E+08		1,34E+09
	8,94E+08		1,33E+09		1,59E+09		1,21E+09
	1,03E+09		1,16E+09		1,21E+09		1,23E+09
582	1,75E+09	583	1,95E+09	585	1,82E+09	586	1,28E+09
	2,14E+09		1,61E+09		1,50E+09		1,29E+09
	1,70E+09		1,11E+09		1,40E+09		1,33E+09
	1,16E+09		5,85E+08		1,35E+09		1,54E+09
	1,01E+09		4,13E+08		1,40E+09		1,35E+09
	1,19E+09		5,43E+08		1,61E+09		1,48E+09
	1,79E+09		1,19E+09		1,68E+09		1,28E+09
	1,49E+09		1,51E+09		1,06E+09		1,34E+09
595	1,14E+09	596	2,26E+09	598	3,16E+09	599	1,82E+09
	2,68E+09		3,27E+09		2,84E+09		1,85E+09
	1,50E+09		3,08E+09		3,43E+09		1,96E+09
	2,45E+09		3,16E+09		2,67E+09		2,23E+09
	2,03E+09		2,71E+09		3,16E+09		2,04E+09
	2,32E+09		2,56E+09		2,29E+09		1,87E+09
	1,41E+09		3,63E+09		2,69E+09		1,85E+09
	2,10E+09		2,13E+09		2,95E+09		1,91E+09
5108	4,23E+09	5109	3,06E+09	5111	2,19E+09	5112	2,61E+09
	3,09E+09		3,31E+09		3,11E+09		3,04E+09
	2,88E+09		2,50E+09		2,67E+09		2,71E+09
	3,34E+09		1,35E+09		2,16E+09		2,56E+09
	2,71E+09		5,00E+08		2,62E+09		2,93E+09
	2,99E+09		2,08E+08		2,95E+09		3,25E+09
	3,89E+09		3,62E+08		2,52E+09		2,88E+09
	3,62E+09		6,19E+08		2,85E+09		2,60E+09

Appendix I

Dynamic stiffnesses T 49.
N/m.

J. J. Rincent, *Anchors—Tests Procedures and Vibratory Analysis*,
https://doi.org/10.1007/978-981-96-3777-5

1	8,51E+07
2	2,10E+06
3	6,13E+07
4	3,03E+06
5	2,30E+06
6	1,88E+06
7	8,74E+06
8	2,81E+05
9	1,56E+08
10	2,07E+08
11	3,37E+08
12	2,20E+08
13	2,46E+08
14	1,96E+08
15	2,07E+08
16	2,47E+08
17	5,40E+07
18	1,32E+08
19	1,09E+08
20	1,74E+08
21	2,79E+08
22	1,58E+08
23	1,39E+08
24	7,61E+07
25	4,40E+06
26	1,15E+07
27	1,58E+07
28	7,86E+05
29	2,52E+06
30	7,60E+06
31	2,15E+05

Appendix J

Dynamic stiffnesses T 238, 3132, 3127, 5125, 5138.

238	2,17E+09		2,17E+09	3132	2,34E+09		2,34E+09
	2,05E+09		2,07E+09		2,57E+09		2,34E+09
	2,07E+09		1,98E+09		2,34E+09		2,54E+09
	1,98E+09		1,95E+09		2,54E+09		2,38E+09
	3,11E+09		2,25E+09		2,38E+09		2,48E+09
	1,95E+09		2,31E+09		2,26E+09		2,46E+09
	2,25E+09				2,48E+09		
	2,31E+09	2,12E+09			2,46E+09	2,42E+09	
		1,46E+08	6,89%			8,24E+07	3,40%

3127	6,12E+08		6,12E+08	5125	2,98E+09		2,98E+09
	8,00E+08		8,00E+08		3,22E+09		3,22E+09
	5,56E+08		5,56E+08		2,51E+09		2,96E+09
	2,14E+09		9,55E+08		2,96E+09		2,91E+09
	9,55E+08		6,55E+08		2,91E+09		3,15E+09
	1,54E+08		7,96E+08		3,15E+09		3,25E+09
	6,55E+08				3,34E+09		
	7,96E+08	7,29E+08			3,25E+09	3,08E+09	
		1,48E+08	20,30%			1,46E+08	4,74%

5138	3,69E+09		5,19E+09
	5,19E+09		5,98E+09
	5,98E+09		5,21E+09
	5,21E+09		5,62E+09
	5,62E+09		6,71E+09
	6,71E+09		4,91E+09
	7,57E+09		
	4,91E+09	5,60E+09	
		6,59E+08	11,77%

J. J. Rincent, *Anchors—Tests Procedures and Vibratory Analysis*,
https://doi.org/10.1007/978-981-96-3777-5

Appendix K

Pylon fixation.

J. J. Rincent, *Anchors—Tests Procedures and Vibratory Analysis*,
https://doi.org/10.1007/978-981-96-3777-5

1	9,63E+08	1,07E+09	7,42E+08	1,40E+09
2	1,13E+09	1,07E+09	7,42E+08	1,40E+09
3	7,29E+08	1,07E+09	7,42E+08	1,40E+09
4	8,36E+08	1,07E+09	7,42E+08	1,40E+09
5	1,13E+09	1,07E+09	7,42E+08	1,40E+09
6	1,66E+09	1,07E+09	7,42E+08	1,40E+09
7	1,33E+09	1,07E+09	7,42E+08	1,40E+09
8	1,19E+09	1,07E+09	7,42E+08	1,40E+09
9	1,13E+09	1,07E+09	7,42E+08	1,40E+09
10	1,17E+09	1,07E+09	7,42E+08	1,40E+09
11	1,43E+09	1,07E+09	7,42E+08	1,40E+09
12	1,52E+09	1,07E+09	7,42E+08	1,40E+09
13	2,06E+09	1,07E+09	7,42E+08	1,40E+09
14	1,96E+09	1,07E+09	7,42E+08	1,40E+09
15	1,11E+09	1,07E+09	7,42E+08	1,40E+09
16	9,55E+08	1,07E+09	7,42E+08	1,40E+09
17	8,92E+08	1,07E+09	7,42E+08	1,40E+09
18	8,82E+08	1,07E+09	7,42E+08	1,40E+09
19	6,58E+08	1,07E+09	7,42E+08	1,40E+09
20	9,58E+08	1,07E+09	7,42E+08	1,40E+09
21	1,36E+09	1,07E+09	7,42E+08	1,40E+09
22	1,60E+09	1,07E+09	7,42E+08	1,40E+09
23	1,33E+09	1,07E+09	7,42E+08	1,40E+09
24	1,48E+09	1,07E+09	7,42E+08	1,40E+09
25	1,36E+09	1,07E+09	7,42E+08	1,40E+09
26	1,31E+09	1,07E+09	7,42E+08	1,40E+09
27	1,62E+09	1,07E+09	7,42E+08	1,40E+09
28	1,89E+09	1,07E+09	7,42E+08	1,40E+09
29	1,15E+09	1,07E+09	7,42E+08	1,40E+09
30	1,09E+09	1,07E+09	7,42E+08	1,40E+09
31	9,03E+08	1,07E+09	7,42E+08	1,40E+09
32	6,45E+08	1,07E+09	7,42E+08	1,40E+09

Appendix L

Micro piles tests stiffnesses values.

J. J. Rincent, *Anchors—Tests Procedures and Vibratory Analysis*,
https://doi.org/10.1007/978-981-96-3777-5

6	3,76E+05		8	1,36E+06	
	1,80E+05			8,72E+06	
	6,64E+05			5,01E+06	
	6,21E+05			3,51E+06	
	6,21E+05			1,65E+07	
	4,56E+05			2,84E+06	
	6,68E+05			1,54E+07	
	3,44E+05			5,56E+06	
	4,09E+05			1,37E+07	
	4,09E+05			5,77E+06	
	3,59E+05			5,46E+06	
	3,59E+05	4,52E+05		6,48E+06	6,99E+06
12	3,76E+05		13	3,92E+06	
	8,32E+05			1,67E+06	
	2,93E+05			6,67E+06	
	1,27E+06			6,22E+06	
	8,93E+05			6,26E+06	
	2,73E+06			1,70E+07	
	2,25E+06			2,13E+07	
	5,00E+05			7,95E+06	
	2,95E+05			5,69E+06	
	2,61E+05			2,46E+07	
	9,76E+05			8,39E+06	
	4,87E+05	5,05E+05		1,10E+07	8,65E+06
31	1,06E+07				
	9,18E+06				
	4,88E+06				
	4,13E+06				
	1,12E+07				
	1,41E+07				
	1,13E+07				
	1,18E+07				
	1,00E+07				
	8,21E+06				
	1,72E+07				
	2,82E+06	9,65E+06			

Appendix M

(19) Europäisches Patentamt / European Patent Office / Office européen des brevets

(11) **EP 1 763 667 B1**

(12) **FASCICULE DE BREVET EUROPEEN**

(45) Date de publication et mention de la délivrance du brevet: **13.02.2008 Bulletin 2008/07**

(21) Numéro de dépôt: **05786101.5**

(22) Date de dépôt: **05.07.2005**

(51) Int Cl.: ***G01N 19/04*** (2006.01) ***G01N 3/30*** (2006.01) ***G01N 3/08*** (2006.01)

(86) Numéro de dépôt international: **PCT/FR2005/001719**

(87) Numéro de publication internationale: **WO 2006/013280 (09.02.2006 Gazette 2006/06)**

(54) **PROCEDE ET DISPOSITIF DE DETERMINATION DE L'EFFORT DE RUPTURE A LA TRACTION D'UN ELEMENT FIXE A UN SUPPORT**

VERFAHREN UND VORRICHTUNG ZUR BESTIMMUNG DER ZUGFESTIGKEIT EINES AN EINEM TRÄGER ANGEBRACHTEN ELEMENTS

METHOD AND DEVICE FOR DETERMINING THE TENSILE BREAKING STRENGTH OF AN ELEMENT FIXED TO A SUPPORT

(84) Etats contractants désignés: **AT BE BG CH CY CZ DE DK EE ES FI FR GB GR HU IE IS IT LI LT LU LV MC NL PL PT RO SE SI SK TR**

(30) Priorité: **06.07.2004 FR 0407467**

(43) Date de publication de la demande: **21.03.2007 Bulletin 2007/12**

(73) Titulaire: **Rincent BTP Services**
91026 Evry Cedex (FR)

(72) Inventeur: **RINCENT, Jean-Jacques,**
RINCENT BTP SERVICES
F-91025 EVRY (FR)

(74) Mandataire: **Intes, Didier Gérard André et al**
Cabinet Beau de Loménie
158, rue de l'Université
75340 Paris Cedex 07 (FR)

(56) Documents cités:
US-A- 3 003 351
US-A- 3 425 263
US-A- 4 542 639
US-A- 4 993 876
US-A- 5 024 091

EP 1 763 667 B1

Il est rappelé que: Dans un délai de neuf mois à compter de la date de publication de la mention de la délivrance du brevet européen, toute personne peut faire opposition au brevet européen délivré, auprès de l'Office européen des brevets. L'opposition doit être formée par écrit et motivée. Elle n'est réputée formée qu'après paiement de la taxe d'opposition. (Art. 99(1) Convention sur le brevet européen).

Printed by Jouve, 75001 PARIS (FR)

J. J. Rincent, *Anchors—Tests Procedures and Vibratory Analysis*,
https://doi.org/10.1007/978-981-96-3777-5

Claims

1. Method for determining the tensile breaking strength (F) of an element (10; 10') fixed to a support (12; 12'), the tensile breaking strength (F) being the minimum tensile stress that, exerted on said element (10; 10'), causes the latter (10; 10') to separate from the support (12; 12'), in which the following steps are carried out:

a) carrying out n cycles each comprising the following steps:

i. putting the element (10; 10') into traction by subjecting the latter (10; 10') to a given tensile stress Fsi,
ii. measuring the deformation Di of the element (10; 10') subjected to this given tensile stress Fsi,
iii. subjecting the element (10; 10') in traction to an impact of a given force Fd in order to generate vibration in the element (10; 10') in traction,
iv. recording the vibratory response (26, 26A, 26B, 28) of the element (10; 10') in traction put into vibration,
v. determining the dynamic stiffness Rdi of the element (10; 10') in traction on the basis of the vibratory response,

b) constructing a stress/deformation curve for the element (10; 10') on the basis of the stress/deformation couples (Fsi, Di), the last point (4) on said curve corresponding to the couple (Fsn,

Dn) for which the tensile stress Fsn applied during the aforementioned cycles (1, 2, 3, 4) is greatest,
c) determining the static stiffnesses Rsi of the element (10; 10') for tensile stresses Fsi other than the stress Fsn, on the basis of a relationship between the trace of the stress/deformation curve and the static stiffness Rsi of the element (10; 10') in traction,
d) defining for said element (10; 10'), a rule (L) of correspondence between the static stiffness Rsi and the dynamic stiffness Rdi, on the basis of the dynamic stiffnesses Rdi and the static stiffnesses Rsi determined for each of the stresses Fsi,
e) determining the static stiffness Rsn for the stress Fsn by applying this rule (L) of correspondence to the dynamic stiffness Rdn determined for this stress Fsn,
f) on the basis of said static stiffness Rsn and the application of said relationship between the trace of the stress/deformation curve and the static stiffness of the element (10; 10') in traction, extrapolating the stress/deformation curve beyond the last point (Fsn, Dn) and determining an asymptote (A) of said extrapolated curve, and
g) defining the tensile breaking strength (F) of the element (10; 10') as being the value of the stress that corresponds to this asymptote (A).

2. Method according to Claim 1, **characterised in that** the relationship between the trace of the stress/deformation curve and the static stiffness Rsi of the element (10; 10') in traction used for step c) consists in considering that the static stiffness Rsi for a point (Fsi, Di) on said curve is the gradient of the tangent to this curve at this point.

3. Method according to Claim 1 or 2, **characterised in that** the rule of correspondence between the dynamic stiffnesses Rdi and the static stiffnesses Rsi defined in step d) is a rule of proportionality.

4. Method according to any one of the preceding claims, **characterised in that** n is comprised between three and five, preferably equal to four.

5. Method according to any one of the preceding claims, **characterised in that** an estimated tensile breaking strength is determined beforehand, in order to choose the given tensile stresses Fsi to be applied to the element (10; 10').

6. Method according to the preceding claim, **characterised in that** the greatest tensile stress Fsn applied during the aforementioned cycles is less than or equal to 70% of the estimated tensile breaking strength.

7. Device for determining the tensile breaking strength of an element (10; 10') fixed to a support (12; 12'), the tensile breaking strength being the minimum tensile stress that, exerted on said element (10; 10'), causes a separation of the latter from the support, including:

- means (14) for putting the element (10; 10') into traction by subjecting the latter to a given tensile stress Fsi,
- means (16) for measuring the deformation Di of the element (10; 10') subjected to this given tensile stress Fsi,
- means (18, 20) for subjecting the element (10; 10') in traction to an impact of a given force Fd, in order to generate vibration in the element (10; 10') in traction,
- means (22) for recording the vibratory response of the element (10; 10') in traction put into vibration,
- means (30) for determining the dynamic stiffness Rdi of the element (10; 10') in traction on the basis of the vibratory response (26, 26A, 26B, 28),
- means (24) for constructing a stress/deformation curve for the element (10; 10') on the basis of the stress/deformation couples (Fsi, Di), the last point on said curve corresponding to the couple (Fsn, Dn) for which the tensile stress Fsn applied during the aforementioned cycles is greatest,
- a memory (M) that contains a ratio between the trace of the stress/deformation curve and the static stiffness Rsi of the element (10; 10') in traction,
- means (32) for determining the static stiffnesses Rsi of the element (10; 10') for tensile stresses Fsi other than the stress Fsn, on the basis of the relationship between the trace of the stress/deformation curve and the static stiffness of the element (10; 10') in traction,
- means (34) for defining and recording, for said element (10; 10'), a rule (L) of correspondence between the static stiffness Rsi and the dynamic stiffness Rdi, on the basis of the dynamic stiffnesses Rdi and of the static stiffnesses Rsi determined for each of the stresses Fsi,
- means (36) for applying this rule (L) of correspondence to the dynamic stiffness Rdn determined for this stress Fsn and thus for determining the static stiffness Rsn for the stress Fsn,
- means (38) for extrapolating the stress/deformation curve beyond the last point (Fsn, Dn) on the basis of said static stiffness Rsn and of said relationship between the trace of the stress/deformation curve and the static stiffness of the element (10; 10') in traction, and means (40) for determining an asymptote (A) of said extrapo-

lated curve, and
- means (42) for recording the value of the tensile breaking strength of the element (10; 10') that corresponds to the value of the stress at this asymptote.

8. Device according to the preceding claim, **characterised in that** it comprises mathematical processing means (T) intended to process the vibratory response (26, 26A, 26B) of the element (10; 10').

Patentansprüche

1. Verfahren zum Bestimmen der Zugfestigkeitskraft (F) eines Elements (10; 10'), das an einem Träger (12;12') angebracht ist, wobei die Zugfestigkeitskraft (F) die Mindestzugkraft ist, die, auf das Element (10; 10') ausgeübt, ein Trennen dieses Letzteren (10;10') von dem Träger (12; 12') verursacht, wobei die folgenden Schritte ausgeführt werden:

a) man führt n Zyklen aus, die jeweils die folgenden Schritte aufweisen:

i. man setzt das Element (10; 10') unter Zug, indem man Letzteres (10; 10') einer gegebenen Zugkraft Fsi aussetzt,
ii. man mißt die Verformung Di des Elements (10; 10'), das dieser gegebenen Zugkraft Fsi ausgesetzt wird,
iii. man setzt das Element (10; 10') unter Zug einem Aufprall mit einer bestimmten Kraft Fd aus, um eine Vibration des Elements (10; 10') unter Zug zu erzeugen,
iv. man mißt die Vibrationsreaktion (26, 26A, 26B, 28) des Elements (10; 10') unter Zug, das zum Schwingen gebracht wird,
v. man bestimmt die dynamische Steifheit Rdi des Elements (10; 10') unter Zug ausgehend von der Vibrationsreaktion,

b) man bildet eine Kraft-/Verformungskurve für das Element (10; 10') ausgehend von den Kraft-/ Verformungsmomenten (Fsi, Di), wobei der letzte Punkt (4) der Kurve dem Moment (Fsn, Dn), für das die im Laufe der oben erwähnten Zyklen (1, 2, 3, 4) angelegte Zugkraft Fsn am größten ist, entspricht,
c) man bestimmt die statischen Steifheiten Rsi des Elements (10; 10') für die anderen Zugkräfte Fsi als die Kraft Fsn, ausgehend von einer Beziehung zwischen dem Verlauf der Kraft-/Verformungskurve und der statischen Steifheit Rsi des Elements (10; 10') unter Zug,
d) man definiert für das Element (10; 10') ein Gesetz (L) der Entsprechung zwischen der statischen Steifheit Rsi und der dynamischen Steifheit Rdi ausgehend von den dynamischen Steifheiten Rdi und den statischen Steifheiten Rsi, die für jede der Kräfte Fsi bestimmt wurden,
e) man bestimmt die statische Steifheit Rs für die Kraft Fsn, indem man dieses Gesetz (L) der Entsprechung an die dynamische Steifheit Rdn, die für diese Kraft Fsn bestimmt wurde, anwendet,
f) ausgehend von der statischen Steifheit Rsn und der Anwendung der Beziehung zwischen dem Verlauf der Kraft-/Verformungskurve und der statischen Steifheit des Elements (10; 10') unter Zug, extrapoliert man die Kraft-/Verformungskurve über den letzten Punkt (Fsn, Dn) hinaus und bestimmt eine Asymptote (A) der extrapolierten Kurve und
g) man definiert die Zugfestigkeitskraft (F) des Elements (10; 10') als den Wert der Kraft, der dieser Asymptote (A) entspricht.

2. Verfahren nach Anspruch 1, **dadurch gekennzeichnet, daß** die Beziehung zwischen dem Verlauf der Kraft-/Verformungskurve und der statischen Steifheit Rsi des Elements (10; 10') unter Zug, die für den Schritt c) verwendet wird, darin besteht, anzunehmen, daß die statische Steifheit Rsi für einen Punkt (Fsi, Di) der Kurve die Neigung der Tangente zu dieser Kurve in diesem Punkt ist.

3. Verfahren nach Anspruch 1 oder 2, **dadurch gekennzeichnet, daß** das Gesetz der Entsprechung zwischen den dynamischen Steifheiten Rdi und den statischen Steifheiten Rsi, das im Schritt d) definiert wird, ein Proportionalitätsgesetz ist.

4. Verfahren nach einem der vorhergehenden Ansprüche, **dadurch gekennzeichnet, daß** n zwischen drei und fünf liegt und vorzugsweise gleich vier ist.

5. Verfahren nach einem der vorhergehenden Ansprüche, **dadurch gekennzeichnet, daß** man vorab eine geschätzte Zugfestigkeitskraft bestimmt, um die gegebenen Zugkräfte Fsi, die an das Element (10; 10') anzulegen sind, auszuwählen.

6. Verfahren nach dem vorhergehenden Anspruch, **dadurch gekennzeichnet, daß** die größte Zugkraft Fsn, die im Laufe der oben genannten Zyklen angelegt wird, kleiner oder gleich 70 % der geschätzten Zugfestigkeitskraft ist.

7. Vorrichtung zum Bestimmen der Zugfestigkeitskraft eines Elements (10; 10'), das an einem Träger (12; 12') befestigt ist, wobei die Zugfestigkeitskraft die Mindestzugkraft ist, die, auf das Element (10; 10') ausgeübt, ein Trennen dieses Letzteren von dem Träger verursacht, folgendes aufweisend:

- Mittel (14), um das Element (10; 10') unter Zug zu setzen, indem Letzteres einer gegebenen Zugkraft Fsi ausgesetzt wird,
- Mittel (16) zum Messen der Verformung Di des Elements (10; 10'), das dieser gegebenen Zugkraft Fsi ausgesetzt wird,
- Mittel (18, 20), um das Element (10; 10') unter Zug einem Aufprall mit bestimmter Kraft Fd auszusetzen, um eine Vibration des Elements (10; 10') unter Zug zu erzeugen,
- Mittel (22) zum Messen der Vibrationsreaktion des Elements (10; 10') unter Zug, das in Schwingung versetzt wurde,
- Mittel (30) zum Bestimmen der dynamischen Steifheit Rdi des Elements (10; 10') unter Zug ausgehend von der Vibrationsreaktion (26, 26A, 26B, 28),
- Mittel (24) zum Aufbauen einer Kraft-/Verformungskurve für das Element (10; 10') ausgehend von den Kraft-/Verformungsmomenten (Fsi, Di), wobei der letzte Punkt der Kurve dem Moment (Fsn, Dn) entspricht, bei dem die Zugkraft Fsn, die im Laufe der oben genannten Zyklen angelegt wird, am größten ist,
- einen Speicher (M), der eine Beziehung zwischen dem Verlauf der Kraft-/Verformungskurve und der statischen Steifheit Rsi des Elements (10; 10') unter Zug enthält,
- Mittel (32) zum Bestimmen der statischen Steifheiten Rsi des Elements (10; 10') für die anderen Zugkräfte Fsi als die Kraft Fsn, ausgehend von der Beziehung zwischen dem Verlauf der Kraft-/Verformungskurve und der statischen Steifheit des Elements (10; 10') unter Zug,
- Mittel (34) zum Definieren und Aufzeichnen für das Element (10; 10') eines Gesetzes (L) der Entsprechung zwischen der statischen Steifheit Rsi und der dynamischen Steifheit Rdi ausgehend von den dynamischen Steifheiten Rdi und statischen Steifheiten Rsi, die für jede der Kräfte Fsi bestimmt werden,
- Mittel (36) zum Anwenden dieses Gesetzes (L) der Entsprechung an die dynamische Steifheit Rdn, die für diese Kraft Fsn bestimmt wird und dadurch Bestimmen der statischen Steifheit Rsn für die Kraft Fsn,
- Mittel (38) zum Extrapolieren der Kraft-/Verformungskurve über den letzten Punkt (Fsn, Dn) hinaus, ausgehend von der statischen Steifheit Rsn und der Beziehung zwischen dem Verlauf der Kraft-/Verformungskurve und der statischen Steifheit des Elements (10; 10') unter Zug, und Mittel (40) zum Bestimmen einer Asymptote (A) der extrapolierten Kurve, und
- Mittel (42) zum Messen des Zugfestigkeitskraftwerts des Elements (10; 10'), der dem Kraftwert bei dieser Asymptote entspricht.

8. Vorrichtung nach dem vorhergehenden Anspruch, **dadurch gekennzeichnet, daß** sie mathematische Verarbeitungsmittel (T) aufweist, die dazu bestimmt sind, die Vibrationsreaktion (26, 26A, 26B) des Elements (10; 10') zu verarbeiten.

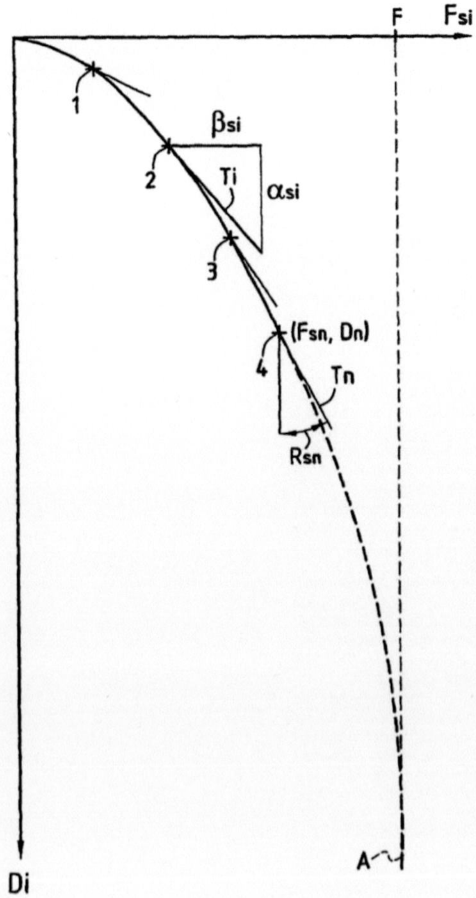
F
Fsi
1
βsi
2
Ti
αsi
3
(Fsn, Dn)
4
Tn
Rsn
A
Di

Appendix N

Dynamic stiffnesses T37.

J. J. Rincent, *Anchors—Tests Procedures and Vibratory Analysis*, https://doi.org/10.1007/978-981-96-3777-5

Force tons	Dynamic stiffness N/m	Average	Square root	Force tons	Dynamic stiffness N/m	Average	Square root
1	7,46E+08			20	1,24E+09		
	8,08E+08				1,22E+09		
	7,00E+08				8,77E+08		
	7,42E+08				1,02E+09		
	7,49E+08				9,13E+08		
	6,57E+08				1,04E+09		
	6,43E+08				9,96E+08		
	7,97E+08	7,32E+08	27055		1,02E+09	1,03E+09	32094
5	6,94E+08			25	1,24E+09		
	7,83E+08				1,06E+09		
	5,75E+08				9,08E+08		
	7,14E+08				8,95E+08		
	7,49E+08				7,96E+08		
	7,22E+08				1,01E+09		
	8,52E+08				9,10E+08		
	8,99E+08	7,52E+08	27423		8,51E+08	9,30E+08	30496
10	2,31E+08			30	1,19E+09		
	2,65E+08				1,06E+09		
	4,48E+08				1,01E+09		
	4,53E+08				9,35E+08		
	6,68E+08				7,30E+08		
	7,38E+08				9,85E+08		
	3,09E+08				1,03E+09		
	6,56E+08	4,67E+08	21610		8,90E+08	9,85E+08	31385
	1,31E+09	15					
	7,61E+08						
	8,12E+08						
	8,45E+08						
	1,15E+09						
	9,17E+08						
	7,76E+08						
	1,11E+09	9,35E+08	30578				

Bibliography

American Concrete Institute Report (1998) ACI 228.2R-98. Non-destructive test methods for evaluation of concrete in structures. ACI, Farmington Hills, Michigan, p 62

Ancrages TA (2020) Règles professionnelles concernant la conception, le calcul, l'exécution, le contrôle et la surveillance

Asli C (2008) Analyse des mesures et combinaison des raideurs dynamiques IUP Evry

Asli C (2013) Thèse: analyse dynamique des structures multicouches soumise à une force d'impact: application aux chaussées. Evry-Val d'Essonne

ASN Autorité de sûreté nucléaire CODEP—DCN 2015–0 17985 Tirants d'ancrage précontraints

ASTRA 120005 Tirants d'ancrage Edition (2022) V1. 13

Borrell Ferreres D (2010). Étude à long terme et à la fatigue des systèmes d'ancrage de pièces métalliques dans le béton

Brevet P, Olivié F, Guilbaud JP, Raharinaivo A (1970–2000) Microstructure et propriétés mécaniques des aciers pour câbles. Synthèse des travaux du LCPC Plasticité et endommagement

Briard M (1970) Contrôle des pieux par la méthode des vibrations. Annales de l'ITBTP Institut Technique du Bâtiment

CFBR Comité français des barrages et réservoirs (2015) Symposium du CFBR à Grenoble Contrôle de la tension des tirants

Davis AG, Dunn CS (1974) From theory to field experience with the non-destructive vibration testing of piles. Proc Instn Civ Eng Port 2(57):571–593

Davis AG, Lim MK, Petersen CG (2005) Rapid and economical evaluation of concrete tunnel linings with impulse response and impulse radar non-destructive methods. NDT & E Int 38(3):181–186

Davis AG, Guillermain P (1979) Interprétation géotechnique des courbes de réponse de l'excitation harmonique d'un pieu. Revue Française de Géotechnique 8 :15–21

Davis AG, Guillermain P (1980) La vibration des pieux: interprétation géotechnique, annales I.T.B.T.P., série SF/170, n° 380, pp. 70–86

Davis AG, Hertlein BH (1995). Non-destructive testing of concrete chimneys and other structures. In: Nazarian S, Olson L (eds) Conference non-destructive evaluation of aging structures and dams, proceedings of SPIE 2457, Oakland, CA, p 129 36

Davis AG, Kennedy J (1998) Impulse response testing to evaluate the degree of alkali-aggregate reaction in concrete drilled shaft foundations under electricity transmission towers. In: Conference on non-destructive evaluation of Utilities and Pipelines, proceedings of SPIE 3398, paper 21, San Antonio, TX

Davis AG, Robertson SA (1976) Vibration testing of piles. Struct Eng A7–Al0

J. J. Rincent, *Anchors—Tests Procedures and Vibratory Analysis*, https://doi.org/10.1007/978-981-96-3777-5

Heeyong H, Heedong G, Jun Won K, Stijn F, Loukas FK (2023) Using the impulse-response pile data for soil characterization

Hertlein BH, Davis AG (1998)) Locating concrete consolidation problems using the non-destructive impulse response test. American Concrete Institute Fall Convention, Los Angeles, CA

Higgs JS, Robertson SA (1979) Integrity testing of concrete piles by shock method. Concrete 13(10):31–33

Holeyman AE (1992) Application of stress-wave theory to piles, F. 8. J Barends (ed) © Balkema, Rotterdam. ISBN 90 5410 082 6 Keynote lecture: Technology of pile dynamic testing Harding Lawson Associates, Santa Ana, Calif., USA

Horb C, Saurel J, Rincent JJ (2021) Preventive maintenance and non-destructive monitoring of rock anchorages Tools resulting from dynamic tests and their correlation with static tests. 5E Stabilité de la pente rocheuse RSS Symposium, Chambéry

Knothe K, Yu M, Illias H (2002) Measurement and modelling of resilient rubber rail-pads system. Dynamics and long-term behaviour of Railway Vehicles Track and Subgrade, Concluding Colloquium of the DFG-Priority-Program Stuttgart

Ladiges S, Wark R, Rodd R (2016) maintenance and testing of post-tensioned anchors for dams and appurtenant structures

Mitaine L, Rincent JJ (2015) Innovative non-destructive technique for determining tension in ground anchors Revue Paralia, vol 8, pp s01.1–s01.7. https://doi.org/10.5150/revue-paralia.2015.s01

Norme suisse SIA 267/1 (2013) Géotechnique

Norme française (1994) NF P 94 160-4 Sols : reconnaissance et essais Auscultation d'un élément de fondations partie 4 méthode par impédance

Ottosen NS, Ristinmaa M, Davis AG (2004) Theoretical interpretation of impulse response tests of embedded concrete structures. J Eng Mech 130(9):1062–1071

Paquet J (1991) Une nouvelle orientation dans le contrôle de l'intégrité des pieux par sollicitation dynamique: Le Profil d'Impédance, Colloque Int. Fondations Profondes, Paris, 10 p

Paquet J (1968) Etude Vibratoire des Pieux en Béton, réponse harmonique et impulsionnelle—application au Contrôle, Annales de l'ITBTP, No. 245, Mai, Série Essais et Mesures 111

Paquet J, Briard M (1976) Contrôle non destructif des pieux en béton—Annales de l'ITBTP n° 337

Paquet J (1988) Checking bearing capacity of dynamic loading—choice of a methodology. In: Proceedings of the 3rd international conference the application of stress wave theory to piles, Ottawa, May, pp 383–398

Piron A, Chulliat O, Morel F (2012) Suivi de la tension résiduelle par auscultation des tirants d'ancrage du parc hydraulique EDF

Rincent JJ (2006) Method and device for determining the tensile breaking strength of an element fixed to a support WO 2006/013280 A1

Rincent JJ (2023) Ensaios não destrutivos Análise vibratória Aplicações em Tirantes—Força. Livro RG EDITORES LTDA

Rincent JJ (2006) Method and device for determining the tensile stress exerted on a sealed element. Patent OMPI n° WO 2006/010830 A1

Rincent JJ (2024) ground anchors tension force—vibratory analysis. Book Open Access Springer © 2024 https://doi.org/10.1007/978-981-97-4414-5

Soyez L (2011) Contribution à l'étude du comportement des ouvrages de soutènement renforcés soumis à des charges d'exploitation ferroviaires. (Charges dynamiques et cycliques)

Stain RT (1982) Integrity testing, civil engineering (London), Apr 1982, pp 53–59 and May 1982, pp 71–73

Turner MJ (1997) Integrity testing in piling practice. In: Construction Industry Research and Information Association (CIRIA) Publication: Ground Engineering, R144. CIRIA, London, UK

Vié D (2004) Journées Nationales de Géotechnique et de Géologie de l'ingénieur, Lille Chargement dynamique des pieux essais à bas niveau d'énergie

Wilquin F, Horb C, Feng ZQ, Porcher G (2016) Dynamic non-destructive evaluation of rock anchorage. In: 3rd International Rock Slope Stability Lyon

MIX
Papier aus verantwortungsvollen Quellen
Paper from responsible sources
FSC® C105338

If you have any concerns about our products,
you can contact us on
ProductSafety@springernature.com

In case Publisher is established outside the EU,
the EU authorized representative is:
Springer Nature Customer Service Center GmbH
Europaplatz 3, 69115 Heidelberg, Germany

Printed by Libri Plureos GmbH
in Hamburg, Germany